Natural Computing Series

Series Editors: G. Rozenberg
Th. Bäck A.E. Eiben J.N. Kok H.P. Spaink

Leiden Center for Natural Computing

For further volumes:
www.springer.com/series/4190

Franz Rothlauf

Design of Modern Heuristics

Principles and Application

 Springer

Prof. Dr. Franz Rothlauf
Chair of Information Systems
  and Business Administration
Johannes Gutenberg Universität Mainz
Gutenberg School of Management
  and Economics
Jakob-Welder-Weg 9
55099 Mainz
Germany
rothlauf@uni-mainz.de

Series Editors

G. Rozenberg (Managing Editor)
rozenber@liacs.nl

Th. Bäck, J.N. Kok, H.P. Spaink
Leiden Center for Natural Computing
Leiden University
Niels Bohrweg 1
2333 CA Leiden, The Netherlands

A.E. Eiben
Vrije Universiteit Amsterdam
The Netherlands

ISSN 1619-7127   Natural Computing Series
ISBN 978-3-642-27070-3   ISBN 978-3-540-72962-4 (eBook)
DOI 10.1007/978-3-540-72962-4
Springer Heidelberg Dordrecht London New York

ACM Computing Classification (1998): I.2.8, G.1.6, H.4.2

Cover design: KünkelLopka GmbH, Heidelberg

Printed on acid-free paper

Springer is part of Springer Science+Business Media (www.springer.com)

Gewidmet Elisabeth[2] und Alfons Rothlauf

Foreword

This is the kind of book that all students interested in search and optimization should read. Almost half of the book is devoted to laying a foundation for understanding search and optimization for both heuristic methods and more traditional exact methods. This is really why the book is so valuable. It puts heuristic search in context, and this integrated view is important and is often lacking in other books on modern heuristic search. The book then goes on to provide an excellent tutorial level discussion of heuristic methods such as evolutionary algorithms, variable neighborhood search, iterated local search and tabu search. The book is a valuable contribution to the field. Other books have tried to provide the same breadth by collecting together tutorials by multiple authors. But Prof. Rothlauf's book is stronger because it provides an integrated and unified explanation of modern heuristic methods.

Darrell Whitley, Colorado State University,
chair of SIGEVO, former Editor-in-Chief of the journal Evolutionary Computation

The book by Franz Rothlauf is interesting in many ways. First, it goes much further than a simple description of the most important modern heuristics; it provides insight into the reasons that explain the success of some methods. Another attractive feature of the book is its thorough, yet concise, treatment of the complete scope of optimization methods, including techniques for continuous optimization; this allows readers with a limited background in optimization to gain a deeper appreciation of the modern heuristics that are the main topic of the book. Finally, the case studies presented provide a nice illustration of the application of modern heuristics to challenging and highly relevant problems.

Michel Gendreau, École Polytechnique de Montréal, former Vice-President of
International Federation of Operational Research Societies (IFORS) and the
Institute for Operations Research and the Management Sciences (INFORMS),
Editor-in-Chief of the journal Transportation Science

Franz Rothlauf's new book, Design of Modern Heuristics: Principles and Application, is a celebration of computer science at its best, combining a blend of mathematical analysis, empirical inquiry, conceptual modeling, and useful application to give readers a principled and practical overview of heuristics in modern problem solving. Despite years of successful experience to the contrary, some still use the word "heuristic" as a putdown or dirty word, suggesting that any computational procedure not formally proven to converge on some particular class of problem is somehow not worthy of study or even polite discussion. Rothlauf's intelligent text sets such folly aside, helping readers to understand that a principled approach can be taken to those computational objects that defy simple mathematical description or elucidation, helping readers to separate the wheat from the computational chaff clearly and systematically to practical end.

David E. Goldberg, University of Illinois at Urbana-Champaign,
author of "Genetic Algorithms in Search, Optimization, and Machine Learning"
and "The Design of Innovation"

The book on modern heuristic methods by Franz Rothlauf is very special as it has very strong practical flavour – it teaches us how to design efficient and effective modern heuristics to solve a particular problem. This emphasis on design component resulted in in-depth discussion on topics like: for which types of problems we should use modern heuristics, how we can select a modern heuristic that fits well to our problem, what are basic principles for the design of modern heuristics, and how we can use problem-specific knowledge for the design of modern heuristics? I recommend this book highly to the whole optimization research community and, in particular, to every practitioner who is interested in applicability of modern heuristic methods.

Zbigniew Michalewicz, University of Adelaide,
author of "How to Solve It: Modern Heuristics"

Contents

Chapter 1
Introduction

Modern heuristics like genetic algorithms, simulated annealing, genetic programming, tabu search, and others are easy-to-apply optimization concepts that are widely used for fully-automated decision making and problem solving. Users who want to apply these powerful methods to solve a particular optimization problem are often confronted with three fundamental challenges:

1. Which optimization method is the *right* optimization method for our problem and can solve it with reasonable effort? Should we try traditional optimization methods like branch-and-bound or cutting plane methods which guarantee finding an optimal solution but have high effort, or should we instead use modern heuristics where we have no guarantee on solution quality but are often able to find reasonable solutions in relatively short time? Finding answers to this question is difficult as the two research areas, classical optimization and modern heuristics, are separate and not much interaction exists. Modern heuristic researchers often argue that once a problem is NP-hard, modern heuristics are the method of choice. However, they are not aware that for many problems also optimization methods are available that are relatively fast and provide guarantees on their solution quality. In contrast, researchers working on classical, non-heuristic optimization methods are often scared by heuristics which do not guarantee finding optimal solutions. They are not aware that modern heuristics can be effective and efficient optimization tools for a large variety of problems and have a solid body of underlying theory.
2. After we have decided that we want or have to use modern heuristics for solving our problem, we want to know which type of modern heuristic we should use. Should we use an existing one from the literature or do we have to design a new one? A look in the literature often does not give us clear answers. We find a vast number of different types of modern heuristics (probably there are more different types of modern heuristics than researchers in this field) but less guidance on what method fits to which type of problem. Many textbooks on modern heuristics provide us with detailed descriptions of the functionality of single modern heuristics, but neglect to teach us the underlying general concepts behind all these different approaches. Therefore, a systematic choice or design of modern

F. Rothlauf, *Design of Modern Heuristics*, Natural Computing Series,
DOI 10.1007/978-3-540-72962-4_1, © Springer-Verlag Berlin Heidelberg 2011

heuristics is difficult. To deliberately decide between different types of modern heuristics, we must have a good understanding of their fundamental principles. Otherwise, their application is a trial-and-error process, where we change and tweak design options and parameters until the method finally delivers solutions of reasonable quality. Such an unsystematic approach is very time-consuming and frustrating if we just want to apply modern heuristics to solve our problems.

3. Finally, after we have decided about the type of modern heuristics, we want to know how we can systematically make use of knowledge about our problem for the design of a modern heuristic. There is a general trade-off between the effectiveness and application range of optimization methods. Usually, the more problems can be solved with one particular optimization method, the lower is its resulting average performance. Therefore, standard modern heuristics that are not problem-specific often only work for small or toy problems. If the problem gets larger and more realistic, performance degrades. To increase the performance for selected optimization problems, we must design modern heuristics in a more problem-specific way. This is possible if we have some idea about properties of good or even bad solutions to our problem. However, we need advice on how to systematically exploit such problem-specific knowledge.

This book addresses these problems and teaches us how to design efficient and effective modern heuristics. With the term modern heuristics, we denote improvement heuristics that can be applied to a wide range of different problems (they are general-purpose methods) and that use during search both intensification (exploitation) and diversification (exploration) steps. Throughout the book, we learn

- for which types of problems we should use modern heuristics,
- how we can select a modern heuristic that fits well to our problem,
- what are common design elements of modern heuristics,
- how we can categorize different types of modern heuristics,
- what are basic principles for the design of modern heuristics, and
- how we can use problem-specific knowledge for the design of modern heuristics.

We do not study single modern heuristics independently of each other, but provide an integrated view by focusing on common design elements and general design principles of modern heuristics. To answer the questions and to make life easier for students and practitioners who want to design and apply modern heuristics, the book is divided into three parts.

1. The first part gives us an overview of different optimization approaches and explains which types of optimization methods fit to which types of optimization problems. Therefore, it does not focus only on modern heuristics but also provides an overview of traditional optimization methods like the Simplex method, branch-and-bound, dynamic programming, cutting plane methods, or approximation algorithms. The goal is not an in-depth analysis of single methods but a basic understanding of method principles allowing users to judge which methods are appropriate for their problem. In combination with a profound understanding of relevant properties of optimization problems, we are able to systematically decide whether modern heuristics are the *right* method for our problem.

2. The second part of the book teaches us how to design efficient and effective modern heuristics. It studies basic design elements (representation, search operator, fitness function, initialization, and search strategy), illustrates relevant design principles (locality and bias), and presents a coherent categorization of modern heuristics. We learn why locality is relevant for the design of representations and search operators and how we can exploit problem-specific knowledge by biasing the design elements.

3. The third part presents two case studies on the systematic design of modern heuristics. In the first one, we examine the locality of two approaches for automated programming and illustrate why high locality is important for successful use of modern heuristics. The second case study is about the design of problem-specific modern heuristics for the optimal communication spanning tree problem. We demonstrate how to make use of problem-specific knowledge for the design of representations, search operators, and initial solutions. In particular, we exploit the bias of optimal solutions towards minimum spanning trees by introducing an analogous bias in the design elements of modern heuristics. The results emphasize that it is not the choice of a particular type of modern heuristic that is relevant for high performance, but an appropriate consideration of problem-specific knowledge.

In detail, the book is structured as follows: Chap. 2 illustrates the process of finding high-quality solutions and discusses relevant properties of optimization problems. We introduce the locality and decomposability of a problem and describe how they affect local and recombination-based search methods. Chapter 3 provides an overview of combinatorial optimization methods. Section 3.2 discusses relevant optimization methods for linear, continuous problems. Such problems are relatively "easy" as they can be solved in polynomial time. Common optimization methods are the Simplex method and interior point methods. Section 3.3 focuses on linear, discrete problems and describes the functionality of selected optimization methods like informed and uninformed search, branch-and-bound, dynamic programming, and cutting plane methods. These methods guarantee returning optimal solutions, however, for NP-hard problems their effort grows exponentially with the problem size. Section 3.4 gives an overview of heuristic optimization methods. It distinguishes between simple construction and improvement heuristics, approximation algorithms, and modern heuristics. Modern heuristics differ from simple improvement heuristics, which only use intensifying elements, by the use of intensifying *and* diversifying elements. Approximation algorithms are heuristics for which bounds on their worst-case performance exist. The chapter ends with the no-free-lunch theorem which tells us that black-box optimization, where an algorithm does not make use of information learned about a particular problem, is not efficient, but high-quality modern heuristics must be problem-specific.

Chapter 4 discusses common and fundamental design elements of modern heuristics, namely representation and search operator, fitness function, initialization, and search strategy. These design elements are relevant for all different types of modern heuristics and understanding the general concepts behind them is a prerequisite for their systematic design. Representations are mappings that assign problem solutions (phenotypes) to, usually linear, strings (genotypes). Given the genotypes, we can define a search space either by defining search operators or by formulating explicit neighborhood relationships between solutions. Solutions that are created by a local search operator are neighbors. Recombination operators generate offspring, where the distances between offspring and parents are usually equal to or smaller than the distance between parents. Therefore, the definition of search operators directly implies a neighborhood structure on the search space. Designing a fitness function and initialization method is usually easier than designing proper representations and operators. The fitness function is determined by the objective function and allows modern heuristics to perform pairwise comparisons between solutions. Initial solutions are usually randomly created if no a priori knowledge about a problem exists.

Chapter 5 presents the fifth design element which is the concept for controlling the search. Search strategies differ in the design and control of intensification and diversification. Diversification steps randomize search by performing large changes of solutions but allow modern heuristics to escape from local optima. Diversification can be introduced into search by a proper design of a representation or search operator, a fitness function, initial solutions, or explicit control of the search strategy. Consequently, we classify modern heuristics according to their diversification mechanisms and present representative examples of local search methods (variable neighborhood search, guided local search, iterated local search, simulated annealing, tabu search, and evolution strategies) and recombination-based search methods (genetic algorithms, estimation of distribution algorithms, and genetic programming).

Chapter 6 presents two general principles for the design of modern heuristics: locality and bias. Assuming that the vast majority of real-world problems are neither deceptive nor (very) difficult and have high locality, modern heuristics must ensure that their design does not destroy the high locality of a problem. Therefore, the search operators used should fit the search space and representations should have high locality: this means similarities between phenotypes must correspond to similarities between genotypes. Second, the performance of modern heuristics can be increased by problem-specific knowledge. Consequently, we study how properties of high-quality solutions can be exploited by biasing representations, search operators, initial solutions, fitness functions, or search strategies.

Chapters 7 and 8 present two case studies on the design of modern heuristics using the design principles outlined in Chap. 6. In Chap. 7, we examine how the locality of a representation affects the performance of modern heuristics. We examine grammatical evolution which is a variant of genetic programming using as genotypes linear strings instead of parse trees. We find that the locality of the representation used in grammatical evolution is lower than for genetic programming, which reduces the performance of local search approaches. In Chap. 8, we study how to consider problem-specific knowledge by biasing representations, search operators, or initial solutions. We find that optimal solutions for the optimal communication spanning tree problem are similar to the minimum spanning tree (MST). Consequently, biasing the representation, operator, or initial solutions such that MST-like solutions are preferred results in efficient and effective modern heuristics. The book ends with a summary of the main findings of this work.

Part I
Fundamentals

Chapter 2
Optimization Problems

Optimization problems are common in many disciplines and various domains. In optimization problems, we have to find solutions which are optimal or near-optimal with respect to some goals. Usually, we are not able to solve problems in one step, but we follow some process which guides us through problem solving. Often, the solution process is separated into different steps which are executed one after the other. Commonly used steps are recognizing and defining problems, constructing and solving models, and evaluating and implementing solutions.

Combinatorial optimization problems are concerned with the efficient allocation of limited resources to meet desired objectives. The decision variables can take values from bounded, discrete sets and additional constraints on basic resources, such as labor, supplies, or capital, restrict the possible alternatives that are considered feasible. Usually, there are many possible alternatives to consider and a goal determines which of these alternatives is best. The situation is different for continuous optimization problems which are concerned with the optimal setting of parameters or continuous decision variables. Here, no limited number of alternatives exist but optimal values for continuous variables have to be determined.

The purpose of this chapter is to set the stage and give an overview of properties of optimization problems that are relevant for modern heuristics. We describe the process of how to create a problem model that can be solved by optimization methods, and what can go wrong during this process. Furthermore, we look at important properties of optimization models. The most important one is how difficult it is to find optimal solutions. For some well-studied problems, we can give upper and lower bounds on problem difficulty. Other relevant properties of optimization problems are their locality and decomposability. The locality of a problem is exploited by local search methods, whereas the decomposability is exploited by recombination-based search methods. Consequently, we discuss the locality and decomposability of a problem and how it affects the performance of modern heuristics.

The chapter is structured as follows: Sect. 2.1 describes the process of solving optimization problems. In Sect. 2.2, we discuss problems and problem instances. Relevant definitions and properties of optimization models are discussed in Sect. 2.3. We describe common metrics that can be defined on a search space, resulting neigh-

F. Rothlauf, *Design of Modern Heuristics*, Natural Computing Series,
DOI 10.1007/978-3-540-72962-4_2, © Springer-Verlag Berlin Heidelberg 2011

borhoods, and the concept of a fitness landscape. Finally, Sect. 2.4 deals with properties of problems. We review complexity theory as a tool for formulating upper and lower bounds on problem difficulty. Furthermore, we study the locality and decomposability of a problem and their importance for local and recombination-based search, respectively.

2.1 Solution Process

Researchers, users, and organizations like companies or public institutions are confronted in their daily life with a large number of planning and optimization problems. In such problems, different decision alternatives exist and a user or an organization has to select one of these. Selecting one of the available alternatives has some impact on the user or the organization which can be measured by some kind of evaluation criteria. Evaluation criteria are selected such that they describe the (expected) impact of choosing one of the different decision alternatives. In optimization problems, users and organizations are interested in choosing the alternative that either maximizes or minimizes an evaluation function which is defined on the selected evaluation criteria.

Usually, users and organizations cannot freely choose from all available decision alternatives but there are constraints that restrict the number of available alternatives. Common restrictions come from law, technical limitations, or interpersonal relations between humans. In summary, optimization problems have the following characteristics:

- Different decision alternatives are available.
- Additional constraints limit the number of available decision alternatives.
- Each decision alternative can have a different effect on the evaluation criteria.
- An evaluation function defined on the decision alternatives describes the effect of the different decision alternatives.

For optimization problems, a decision alternative should be chosen that considers all available constraints and maximizes/minimizes the evaluation function. For planning problems, a rational, goal-oriented planning process should be used that systematically selects one of the available decision alternatives. Therefore, planning describes the process of generating and comparing different courses of action and then choosing one prior to action.

Planning processes to solve planning or optimization problems have been of major interest in operations research (OR) (Taha, 2002; Hillier and Lieberman, 2002; Domschke and Drexl, 2005). Planning is viewed as a systematic, rational, and theory-guided process to analyze and solve planning and optimization problems. The planning process consists of several steps:

1. Recognizing the problem,
2. defining the problem,
3. constructing a model for the problem,

4. solving the model,
5. validating the obtained solutions, and
6. implementing one solution.

The following sections discuss the different process steps in detail.

2.1.1 Recognizing Problems

In the very first step, it must be recognized that there is a planning or optimization problem. This is probably the most difficult step as users or institutions often quickly get used to a currently used approach of doing business. They appreciate the current situation and are not aware that there are many different ways to do their business or to organize a task. Users or institutions are often not aware that there might be more than one alternative they can choose from.

The first step in problem recognition is that users or institutions become aware that there are different alternatives (for example using a new technology or organizing the current business in a different way). Such an analysis of the existing situation often occurs as a result of external pressure or changes in the environment. If everything goes well, users and companies do not question the currently chosen decision alternatives. However, when running into economic problems (for example accumulating losses or losing market share), companies have to think about re-structuring their processes or re-shaping their businesses. Usually, a re-design of business processes is done with respect to some goals. Designing the proper (optimal) structure of the business processes is an optimization problem.

A problem has been recognized if users or institutions have realized that there are other alternatives and that selecting from these alternatives affects their business. Often, problem recognizing is the most difficult step as users or institutions have to abandon the current way of doing business and accept that there are other (and perhaps better) ways.

2.1.2 Defining Problems

After we have identified a problem, we can describe and define it. For this purpose, we must formulate the different decision alternatives, study whether there are any additional constraints that must be considered, select evaluation criteria which are affected by choosing different alternatives, and determine what are the goals of the planning process. Usually, there is not only one possible goal but we have to choose from a variety of different goals. Possible goals of a planning or optimization process are either to find an optimal solution for the problem or to find a solution that is better than some predefined threshold (for example the current solution).

An important aspect of problem definition is the selection of relevant decision alternatives. There is a trade-off between the number of decision alternatives and the difficulty of the resulting problem. The more decision alternatives we have to consider, the more difficult it is to choose a proper alternative. In principle, we can consider all possible decision alternatives (independently of whether they are relevant for the problem, or not) and try to solve the resulting optimization problem. However, since such problems can not be solved in a reasonable way, usually only decision alternatives are considered that are relevant and which affect the evaluation criteria. All aspects that have no direct impact on the goal of the planning process are neglected. Therefore, we have to focus on carefully selected parts of the overall problem and find the right level of abstraction.

It is important to define the problem large enough to ensure that solving the problem yields some benefits and small enough to be able to solve the problem. The resulting problem definition is often a simplified problem description.

2.1.3 Constructing Models

In this step, we construct a *model* of the problem which represents its essence. Therefore, a model is a (usually simplified) representative of the real world. Mathematical models describe reality by extracting the most relevant relationships and properties of a problem and formulating them using mathematical symbols and expressions. Therefore, when constructing a model, there are always aspects of reality that are idealized or neglected. We want to give an example. In classical mechanics, the energy E of a moving object can be calculated as $E = \frac{1}{2}mv^2$, where m is the object's mass and v its velocity. This model describes the energy of an object well if v is much lower than the speed of light c ($v \ll c$) but it becomes inaccurate for $v \to c$. Then, other models based on the special theory of relativity are necessary. This example illustrates that the model used is always a representation of the real world.

When formulating a model for optimization problems, the different decision alternatives are usually described by using a set of *decision variables* $\{x_1, \ldots, x_n\}$. The use of decision variables allows modeling of the different alternatives that can be chosen. For example, if somebody can choose between two decision alternatives, a possible decision variable would be $x \in \{0, 1\}$, where $x = 0$ represents the first alternative and $x = 1$ represents the second one. Usually, more than one decision variable is used to model different decision alternatives (for choosing proper decision variables see Sect. 2.3.2). Restrictions that hold for the different decision variables can be expressed by constraints. Representative examples are relationships between different decision variables (e.g. $x_1 + x_2 \le 2$). The *objective function* assigns an *objective value* to each possible decision alternative and measures the quality of the different alternatives (e.g. $f(x) = 2x_1 + 4x_2^2$). One possible decision alternative, which is represented by different values for the decision variables, is called a *solution* of a problem.

To construct a model with an appropriate level of abstraction is a difficult task (Schneeweiß, 2003). Often, we start with a realistic but unsolvable problem model and then iteratively simplify the model until it can be solved by existing optimization methods. There is a basic trade-off between the ability of optimization methods to solve a model (*tractability*) and the similarity between the model and the underlying real-world problem (*validity*). A step-wise simplification of a model by iteratively neglecting some properties of the real-world problem makes the model easier to solve and more tractable but reduces the relevance of the model.

Often, a model is chosen such that it can be solved by using existing optimization approaches. This especially holds for classical optimization methods like the Simplex method or branch-and-bound-techniques which guarantee finding the optimal solution. In contrast, the use of modern heuristics allow us to reduce the gap between reality and model and to solve more relevant problem models. However, we have to pay a price since such methods often find good solutions but we have no guarantee that the solutions found are optimal.

Two other relevant aspects of model construction are the availability of relevant data and the testing of the resulting model. For most problems, is it not sufficient to describe the decision variables, the relationships between the decision variables, and the structure of the evaluation function, but additional parameters are necessary. These parameters are often not easily accessible and have to be determined by using simulation and other predictive techniques. An example problem is assigning jobs to different agents. Relevant for the objective value of an assignment is the order of jobs. To be able to compare the duration of different assignments (each specific assignment is a possible decision alternative), parameters like the duration of one work step, the time that is necessary to transfer a job to a different agent, or the setup times of the different agents are relevant. These additional parameters can be determined by analyzing or simulating real-world processes.

Finally, a model is available which should be a representative of the real problem but is usually idealized and simplified in comparison to the real problem. Before continuing with this model, we must ensure that the model is a valid representative of the real world and really represents what we originally wanted to model. A proper criterion for judging the correctness of a model is whether different decision alternatives are modeled with sufficient accuracy and lead to the expected results. Often, the relevance of a model is evaluated by examining the relative differences of the objective values resulting from different decision alternatives.

2.1.4 Solving Models

After we have defined a model of the original problem, the model can be solved by some kind of *algorithm* (usually an optimization algorithm). An algorithm is a procedure (a finite set of well-defined instructions) for accomplishing some task. An algorithm starts in an initial state and terminates in a defined end-state. The concept of an algorithm was formalized by Turing (1936) and Church (1936) and is at the core of computers and computer science. In optimization, the goal of an algorithm is to find a solution (either specific values for the decision variables or one specific decision alternative) with minimal or maximal evaluation value.

Practitioners sometimes view solving a model as simple, as the outcome of the model construction step is already a model that can be solved by some kind of optimization method. Often, they are not aware that the effort to solve a model can be high and only small problem instances can be solved with reasonable effort. They believe that solving a model is just applying a *black-box optimization method* to the problem at hand. An algorithm is called a black-box algorithm if it can be used without any further problem-specific adjustments.

However, we have a trade-off between tractability and specificity of optimization methods. If optimization methods are to perform well for the problem at hand, they usually need to be adapted to the problem. This is typical for modern heuristics but also holds for classical optimization methods like branch-and-bound approaches. Modern heuristics can easily be applied to problems that are very realistic and near to real-world problems but usually do not guarantee finding an optimal solution. Modern heuristics should not be applied out of the box as black-box optimization algorithms but adapted to the problem at hand. To design high-quality heuristics is an art as they are problem-specific and exploit properties of the model.

Comparing classical OR methods like the Simplex method with modern heuristics reveals that for classical methods constructing a valid model of the real problem is demanding and needs the designer's intuition. Model solution is simple as existing algorithms can be used which yield optimal solutions (Ackoff, 1973). The situation is different for modern heuristics, where formulating a model is often a relatively simple step as modern heuristics can also be applied to models that are close to the real world. However, model solution is difficult as standard variants of modern heuristics usually show limited performance and only problem-specific and model-specific variants yield high-quality solutions (Droste and Wiesmann, 2002; Puchta and Gottlieb, 2002; Bonissone et al, 2006).

2.1.5 Validating Solutions

After finding optimal or near-optimal solutions, we have to evaluate them. Often, a *sensitivity analysis* is performed which studies how the optimal solution depends on variations of the model (for example using different parameters). The use of a sensitivity analysis is necessary to ensure that slight changes of the problem, model, or model parameters do not result in large changes in the resulting optimal solution.

Another possibility is to perform *retrospective tests*. Such tests use historical data and measure how well the model and the resulting solution would have performed if they had been used in the past. Retrospective tests can be used to validate the solutions, to evaluate the expected gains from new solutions, and to identify problems of the underlying model. For the validation of solutions, we must also consider that the variables that are used as input for the model are often based on historical data. In general, we have no guarantee that past behavior will correctly forecast future behavior. A representative example is the prediction of stock indexes. Usually, prediction methods are designed such that they well predict historical developments of stock indexes. However, as the variables that influence stock indexes are continuously changing, accurate predictions of future developments with unforeseen events are very difficult, if not impossible.

2.1.6 Implementing Solutions

Validated solutions have to be implemented. There are two possibilities: First, a validated solution is implemented only once. The outcome of the planning or optimization process is a solution that usually replaces an existing, inferior, solution. The solution is implemented by establishing the new solution. After establishing the new solution, the planning process is finished. An example is the redesign of a company's distribution center. The solution is a new design of the processes in the distribution center. After establishing the new design, the process terminates.

Second, the model is used and solved repeatedly. Then, we have to install a well-documented system that allows the users to continuously apply the planning process. The system includes the model, the solution algorithm, and procedures for implementation. An example is a system for finding optimal routes for deliveries and pick-ups of trucks. Since the problem continuously changes (there are different customers, loads, trucks), we must continuously determine high-quality solutions and are not satisfied with a one-time solution.

We can distinguish between two types of systems. Automatic systems run in the background and no user-interaction is necessary for the planning process. In contrast, *decision support systems* determine proper solutions for a problem and present the solution to the user. Then, the user is free to modify the proposed solution and to decide.

This book focuses on the design of modern heuristics. Therefore, defining and solving a model are of special interest. Although the steps before and afterwards in the process are of equal (or even higher) importance, we refrain from studying them in detail but refer the interested reader to other literature (Turban et al, 2004; Power, 2002; Evans, 2006).

2.2 Problem Instances

We have seen in the previous section how the construction of a model is embedded in the solution process. When building a model, we can represent different decision alternatives using a vector $\mathbf{x} = (x_1, \ldots, x_n)$ of n decision variables. We denote an assignment of specific values to $\mathbf{x}$ as a solution. All solutions together form a set X of solutions, where $\mathbf{x} \in X$.

Decision variables can be either continuous ($\mathbf{x} \in \mathbb{R}^n$) or discrete ($\mathbf{x} \in \mathbb{Z}^n$). Consequently, optimization models are either continuous where all decision variables are real numbers, combinatorial where the decision variables are from a finite, discrete set, or mixed where some decision variables are real and some are discrete. The focus of this book is on models for combinatorial optimization problems. Typical sets of solutions used for combinatorial optimization models are integers, permutations, sets, or graphs.

We have seen in Sect. 2.1 that it is important to carefully distinguish between an optimization problem and possible optimization models which are more or less accurate representations of the underlying optimization problem. However, in optimization literature, this distinction is not consistently made and often models are denoted as problems. A representative example is the traveling salesman problem which in most cases denotes a problem model and not the underlying problem. We follow this convention throughout this book and usually talk about problems meaning the underlying model of the problem and, if no confusion can occur, we do not explicitly distinguish between problem and model.

We want to distinguish between problems and *problem instances*. An instance of a problem is a pair (X, f), where X is a set of feasible solutions $x \in X$ and $f : X \to \mathbb{R}$ is an evaluation function that assigns a real value to every element x of the search space. A solution is feasible if it satisfies all constraints. The problem is to find an $x^* \in X$ for which

$$f(x^*) \geq f(x) \quad \text{for all } x \in X \quad \text{(maximization problem)} \tag{2.1}$$
$$f(x^*) \leq f(x) \quad \text{for all } x \in X \quad \text{(minimization problem),} \tag{2.2}$$

x^* is called a globally optimal solution (or optimal solution if no confusion can occur) to the given problem instance.

An *optimization problem* is defined as a set I of instances of a problem. A problem instance is a concrete realization of an optimization problem and an optimization problem can be viewed as a collection of problem instances with the same properties and which are generated in a similar way. Most users of optimization methods are usually dealing with problem instances as they want to have a better solution for a particular problem instance. Users can obtain a solution for a problem instance since all parameters are usually available. Therefore, it is also possible to compare the quality of different solutions for problem instances by evaluating them using the evaluation function f.

(2.1) and (2.2) are examples of definitions of optimization problems. However, it is expensive to list all possible $x \in X$ and to define the evaluation function f separately for each x. A more elegant way is to use standardized model formulations. A representative formulation for optimization models that is understood by people working with optimization problems as well as computer software is:

$$\text{minimize} \quad z = f(x), \tag{2.3}$$
$$\text{subject to}$$
$$g_i(x) \geq 0, \quad i \in \{1, \ldots, m\},$$
$$h_i(x) = 0, \quad i \in \{1, \ldots, p\},$$
$$x \in W_1 \times W_2 \times \ldots \times W_n, \quad W_i \in \{\mathbb{R}, \mathbb{Z}, \mathbb{B}\}, \quad i \in \{1, \ldots, n\},$$

where x is a vector of n decision variables $x_1, \ldots, x_n$, $f(x)$ is the objective function that is used to evaluate different solutions, and $g(x)$ and $h(x)$ are inequality and equality constraints on the variables x_i. $\mathbb{B}$ indicates the set of binary values $\{0, 1\}$.

By using such a formulation, we can model optimization problems with inequality and equality constraints. We cannot describe models with other types of constraints or where the evaluation function f or the constraints g_i and h_j cannot be formulated in an algorithmic way. Also not possible are multi-criteria optimization problems where more than one evaluation criterion exists (Deb, 2001; Coello Coello et al, 2007; Collette and Siarry, 2004). To formulate models in a standard way allow us to easily recognize relevant structures of the model, to easily add details of the model (e.g. additional constraints), and to feed it directly into computer programs (problem solvers) that can compute optimal or good solutions.

We see that in standard optimization problems, there are decision alternatives, restrictions on the decision alternatives, and an evaluation function. Generally, the decision alternatives are modeled as a vector of variables. These variables are used to construct the restrictions and the objective criteria as a mathematical function. By formulating them, we get a mathematical model relating the variables, constraints, and objective function. Solving this model yields the values of the decision variables that optimize (maximize or minimize) values of the objective function while satisfying all constraints. The resulting solution is referred to as an optimal feasible solution.

2.3 Search Spaces

In optimization models, a search space X is implicitly defined by the definition of the decision variables $x \in X$. This section defines important aspects of search spaces. Section 2.3.1 introduces metrics that can be used for measuring similarities between solutions in metric search spaces. In Sect. 2.3.2, neighborhoods in a search space are defined based on the metric used. Finally, Sects. 2.3.3 and 2.3.4 introduce fitness landscapes and discuss differences between locally and globally optimal solutions.

2.3.1 Metrics

To formulate an optimization model, we need to define a *search space*. Intuitively, a search space contains the set of feasible solutions of an optimization problem. Furthermore, a search space can define relationships (for example distances) between solutions.

Very generally, a search space can be defined as a *topological space*. A topological space is a generalization of metric search spaces (as well as other types of search spaces) and describes similarities between solutions not by defining distances between solutions but by relationships between sets of solutions. A topological space is an ordered pair (X,T), where X is a set of solutions (points) and T is a collection of subsets of X called open sets. A set Y is in X (denoted $Y \subseteq X$) if every element $x \in Y$ is also in X ($x \in Y \Rightarrow x \in X$). A topological space (X,T) has the following properties

1. the empty set $\emptyset$ and whole space X are in T,
2. the intersection of two elements of T is again in T, and
3. the union of an arbitrary number of elements of T is again in T.

We can define different topologies (search spaces) by combining X with different T. For a given X, the most simple search space is $T = \{\emptyset, X\}$, which is called the *trivial topology* or indiscrete topology. The trivial topology can be used for describing search spaces where no useful metrics between the different decision alternatives are known or can be defined. For the definition of a topological space, we need no definition of similarity between the different elements in the search space but the definition of relationships between different subsets is sufficient. We give examples. Given $X = \{a, b, c\}$, we can define the trivial topology with only two subsets $T = \{\{\}, \{a, b, c\}\}$. More complex topologies can be defined by different T. For example, defining four subsets $T = \{\{\}, \{a\}, \{a, b\}, \{a, b, c\}\}$ results in a different topology. For more information on topological spaces, we refer to Buskes and van Rooij (1997) and Bredon (1993).

Metric search spaces are a specialized form of topological spaces where the similarities between solutions are measured by a distance. Therefore, in metric search spaces, we have a set X of solutions and a real-valued distance function (also called a metric)

$$d : X \times X \rightarrow \mathbb{R}$$

that assigns a real-valued distance to any combination of two elements $x, y \in X$. In metric search spaces, the following properties must hold:

$$d(x, y) \geq 0,$$
$$d(x, x) = 0,$$
$$d(x, y) = d(y, x),$$
$$d(x, z) \leq d(x, y) + d(y, z),$$

where $x, y, z \in X$.

An example of a metric space is the set of real numbers $\mathbb{R}$. Here, a metric can be defined by $d(x,y) := |x-y|$. Therefore, the distance between any solutions $x, y \in \mathbb{R}$ is just the absolute value of their differences. Extending this definition to 2-dimensional spaces $\mathbb{R}^2$, we get the *city-block metric* (also known as taxicab metric or Manhattan distance). It is defined for 2-dimensional spaces as

$$d(x,y) := |x_1 - y_1| + |x_2 - y_2|, \tag{2.4}$$

where $x = (x_1, x_2)$ and $y = (y_1, y_2)$. This metric is named the city-block metric as it describes the distance between two points on a 2-dimensional plane in a city like Manhattan or Mannheim with a rectangular ground plan. On n-dimensional search spaces $\mathbb{R}^n$, the city-block metric becomes

$$d(x,y) := \sum_{i=1}^{n} |x_i - y_i|, \tag{2.5}$$

where $x, y \in \mathbb{R}^n$.

Another example of a metric that can be defined on $\mathbb{R}^n$ is the *Euclidean metric*. In Euclidean spaces, a solution $x = (x_1, \ldots, x_n)$ is a vector of continuous values $(x_i \in \mathbb{R})$. The Euclidean distance between two solutions x and y is defined as

$$d(x,y) := \sqrt{\sum_{i=1}^{n} (x_i - y_i)^2}. \tag{2.6}$$

For $n = 1$, the Euclidean metric coincides with the city-block metric. For $n = 2$, we have a standard 2-dimensional search space and the distance between two elements $x, y \in \mathbb{R}^2$ is just a direct line between two points on a 2-dimensional plane.

If we assume that we have a binary space $(x \in \{0,1\}^n)$, a commonly used metric is the binary *Hamming metric* (Hamming, 1980)

$$d(x,y) = \sum_{i=1}^{n} |x_i - y_i|, \tag{2.7}$$

where $d(x,y) \in \{0, \ldots, n\}$. The binary Hamming distance between two binary vectors x and y of length n is just the number of binary decision variables on which x and y differ. It can be extended to continuous and discrete decision variables:

$$d(x,y) = \sum_{i=1}^{n} z_i, \tag{2.8}$$

where

$$z_i = \begin{cases} 0, & \text{for } x_i = y_i, \\ 1, & \text{for } x_i \neq y_i. \end{cases}$$

In general, the Hamming distance measures the number of decision variables on which x and y differ.

2.3.2 Neighborhoods

The definition of a *neighborhood* is important for optimization problems as it determines which solutions are similar to each other. A neighborhood is a mapping

$$N(x) : X \to 2^X, \tag{2.9}$$

where X is the search space containing all possible solutions to the problem. 2^X stands for the set of all possible subsets of X and N is a mapping that assigns to each element $x \in X$ a set of elements $y \in X$. A neighborhood definition can be viewed as a mapping that assigns to each solution $x \in X$ a set of solutions y that are neighbors of x. Usually, the neighborhood $N(x)$ defines a set of solutions y which are in some sense similar to x.

The definition of a topological space (X,T) already defines a neighborhood as it introduces an abstract structure of space in the set X. Given a topological space (X,T), a subset N of X is a neighborhood of a point $x \in X$ if N contains an open set $U \subset T$ containing the point x. We want to give examples. For the trivial topology $(\{a,b,c\}, \{\{\}, \{a,b,c\}\})$, the points in the search space cannot be distinguished by topological means and either all or no points are neighbors to each other. For $(\{a,b,c\}, \{\{\}, \{a\}, \{a,b\}, \{a,b,c\}\})$, the points a and b are neighbors.

Many optimization models use metric search spaces. A metric search space is a topological space where a metric between the elements of the set X is defined. Therefore, we can define similarities between solutions based on the distance d. Given a metric search space, we can use balls to define a neighborhood. For $x \in X$, an (open) ball around x of radius ε is defined as the set

$$B_\varepsilon := \{y \in X | d(x,y) < \varepsilon\}.$$

The ε-neighborhood of a point $x \in X$ is the open set consisting of all points whose distance from x is less than ε. This means that all solutions $y \in X$ whose distance d from x is lower than ε are neighbors of x. By using balls we can define a neighborhood function $N(x)$. Such a function defines for each x a set of solutions similar to x.

Figure 2.1 illustrates the definition of a neighborhood in a 2-dimensional continuous search space $\mathbb{R}^2$ for Euclidean distances (Fig. 2.1(a)) and Manhattan distances (Fig. 2.1(b)). Using an open ball, all solutions y where $d(x,y) < \varepsilon$ are neighboring solutions to x. For Euclidean distances, we use $d(x,y) := \sqrt{(x_1 - y_1)^2 + (x_2 - y_2)^2}$ and neighboring solutions are all solutions that can be found inside of a circle around x with radius ε. For city-block distances, we use $d(x,y) := |x_1 - y_1| + |x_2 - y_2|$ and all solutions inside a rhombus with the vertices $(x_1 - \varepsilon, y_1), (x_1, y_1 + \varepsilon), (x_1 + \varepsilon, y_1), (x_1, y_1 - \varepsilon)$ are neighboring solutions.

It is problematic to apply metric search spaces to problems where no meaningful similarities between different decision alternatives can be defined or do not exist. For such problems, the only option is to define a trivial topology (see p. 16) which assumes that all solutions are neighbors and no meaningful structure on the search

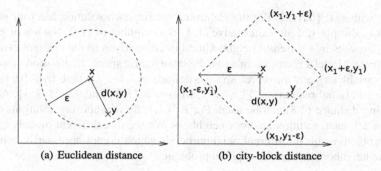

Fig. 2.1 Neighborhoods on a 2-dimensional Euclidean space using different metrics

space exists. However, practitioners (as well as users) are used to metric search spaces and often seek to apply them also to problems where no meaningful similarities between different decision alternatives can be defined. This is a mistake as a metric search space does not model such a problem in an appropriate way.

We want to give an example. We assume a search space containing four different fruits (apple (a), banana (b), pear (p), and orange (o)). This search space forms a trivial topology ($\{a,b,p,o\},\{\emptyset,\{a,b,p,o\}\}$ as no meaningful distances between the four fruits exist and all solutions are neighbors of each other. Nevertheless, we can define a metric search space $X = \{0,1\}^2$ for the problem. Each solution $((0,0),(0,1),(1,0),$ and $(1,1))$ represents one fruit. Although the original problem defines no similarities, the use of a metric space induces that the solution $(0,0)$ is more similar to $(0,1)$ than to $(1,1)$ (using Hamming distance (2.7)). Therefore, a metric space is inappropriate for the problem definition as it defines similarities where no similarities exist. A more appropriate model would be $x \in \{0,\ldots,3\}$ and using Hamming distance (2.8). Then, all distances between the different solutions are equal and all solutions are neighbors.

A different problem can occur if the metric used is not appropriate for the problem and existing "similarities" between different decision alternatives do not fit the similarities between different solutions described by the model. The metric defined for the problem model is a result of the choice of the decision variables. Any choice of decision variables $\{x_1,\ldots,x_n\}$ allows the definition of a metric space and, thus, defines similarities between different solutions. However, if the metric induced by the use of the decision variables does not fit the metric of the problem description, the problem model is inappropriate.

Table 2.1 illustrates this situation. We assume that there are $s = 9$ different decision alternatives $\{a,b,c,d,e,f,g,h,i\}$. We assume that the decision alternatives form a metric space (using the Hamming metric (2.8)), where the distances between all elements are equal. Therefore, all decision alternatives are neighbors (for $\varepsilon > 1$). In the first problem model (model 1), we use a metric space $X = \{0,1,2\}^2$ and Hamming metric (2.8). Therefore, each decision alternative is represented by

(x_1, x_2) with $x_i \in \{0,1,2\}$. For the Hamming metric, each solution has four neighbors. For example, decision alternative $(1,1)$ is a neighbor of $(1,2)$ but not of $(2,2)$. Model 1 results in a different neighborhood in comparison to the original decision alternatives. Model 2 is an example of a different metric space. In this model, we use binary variables x_{ij} and the search space is defined as $X = x_{ij}$, where $x_{ij} \in \{0,1\}$. We have an additional restriction, $\sum_j x_{ij} = 1$, where $i \in \{1,2\}$ and $j \in \{1,2,3\}$. Again, Hamming distance (2.8) can be used. For $\varepsilon = 1.1$, no neighboring solutions exist. For $\varepsilon = 2.1$, each solution has four neighbors. We see that different models for the same problem result in different neighborhoods which do not necessarily coincide with the neighborhoods of the original problem.

Table 2.1 Two different search spaces for a problem

decision alternatives	model 1 (x_1, x_2)	model 2 $\begin{pmatrix} x_{11} & x_{12} & x_{13} \\ x_{21} & x_{22} & x_{23} \end{pmatrix}$
$\{a,b,c,d,e,f,g,h,i\}$	$\{(0,0),(0,1),(0,2),$ $(1,0),(1,1),(1,2),$ $(2,0),(2,1),(2,2)\}$	$\{ \begin{pmatrix} 1 & 0 & 0 \\ 1 & 0 & 0 \end{pmatrix}, \begin{pmatrix} 1 & 0 & 0 \\ 0 & 1 & 0 \end{pmatrix}, \begin{pmatrix} 1 & 0 & 0 \\ 0 & 0 & 1 \end{pmatrix},$ $\begin{pmatrix} 0 & 1 & 0 \\ 1 & 0 & 0 \end{pmatrix}, \begin{pmatrix} 0 & 1 & 0 \\ 0 & 1 & 0 \end{pmatrix}, \begin{pmatrix} 0 & 1 & 0 \\ 0 & 0 & 1 \end{pmatrix},$ $\begin{pmatrix} 0 & 0 & 1 \\ 1 & 0 & 0 \end{pmatrix}, \begin{pmatrix} 0 & 0 & 1 \\ 0 & 1 & 0 \end{pmatrix}, \begin{pmatrix} 0 & 0 & 1 \\ 0 & 0 & 1 \end{pmatrix} \}$

The examples illustrate that selecting an appropriate model is important. For the same problem, different models are possible which can result in different neighborhoods. We must select the model such that the metric induced by the model fits well the metric that exists for the decision alternatives. Although, in the problem description no neighborhood needs to be defined, usually a notion of neighborhoods exist. Users that formulate a model description often know which decision alternatives are similar to each other as they have a feeling about which decision alternatives result in the same outcome. When constructing the model, we must ensure that the neighborhood induced by the model fits well the (often intuitive) neighborhood formulated by the user.

Relevant aspects which determine the resulting neighborhood of a model are the type and number of decision variables. The types of decision variables should be determined by the properties of the decision alternatives. If decision alternatives are continuous (for example, choosing the right amount of crushed ice for a drink), the use of continuous decision variables in the model is useful and discrete decision variables should not be used. Analogously, for discrete decision alternatives, discrete decision variables and combinatorial models should be preferred. For example, the number of ice cubes in a drink should be modelled using integers and not continuous variables.

The number of decision variables used in the model also affects the resulting neighborhood. For discrete models, there are two extremes: first, we can model s different decision alternatives by using only one decision variable that can take s

different values. Second, we can use $l = \log_2(s)$ binary decision variables $x_i \in \{0,1\}$ ($i \in \{1,\ldots,l\}$). If we use Hamming distance (2.8) and define neighboring solutions as $d(x,y) \leq 1$, then all possible solutions are neighbors if we use only one decision variable. In contrast, each solution $x \in \{0,1\}^l$ has only l neighbors $y \in \{0,1\}^l$, where $d(x,y) \leq 1$. We see that using different numbers of decision variables for modeling the decision alternatives results in completely different neighborhoods. In general, a high-quality model is a model where the neighborhoods defined in the model fit well the neighborhoods that exist in the problem.

2.3.3 Fitness Landscapes

For combinatorial search spaces where a metric is defined, we can introduce the concept of *fitness landscape* (Wright, 1932). A fitness landscape (X,f,d) of a problem instance consists of a set of solutions $x \in X$, an objective function f that measures the quality of each solution, and a distance measure d. Figure 2.2 is an example of a one-dimensional fitness landscape.

We denote $d_{min} = \min_{x,y \in X}(d(x,y))$, where $x \neq y$, as the minimum distance between any two elements x and y of a search space. Two solutions x and y are denoted as neighbors if $d(x,y) = d_{min}$. Often, d can be normalized to $d_{min} = 1$. A fitness landscape can be described using a graph G_L with a vertex set $V = X$ and an edge set $E = \{(x,y) \in X \times X \mid d(x,y) = d_{min}\}$ (Reeves, 1999a; Merz and Freisleben, 2000b). The objective function assigns an objective value to each vertex. We assume that each solution has at least one neighbor and the resulting graph is connected. Therefore, an edge exists between neighboring solutions. The distance between two solutions $x,y \in X$ is proportional to the number of nodes that are on the path of minimal length between x and y in the graph G_L. The maximum distance $d_{max} = \max_{x,y \in X}(d(x,y))$ between any two solutions $x,y \in X$ is called the diameter *diam* G_L of the landscape.

We want to give an example: We use the search space defined by model 1 in Table 2.1 and Hamming distance (2.8). Then, all solutions where only one decision variable is different are neighboring solutions ($d(x,y) = d_{min}$). The maximum distance $d_{max} = 2$. More details on fitness landscapes can be found in Reeves and Rowe (2003, Chap. 9) or Deb et al (1997).

2.3.4 Optimal Solutions

A globally optimal solution for an optimization problem is defined as the solution $x^* \in X$, where $f(x^*) \leq f(x)$ for all $x \in X$ (minimization problem). For the definition of a globally optimal solution, it is not necessary to define the structure of the search space, a metric, or a neighborhood.

Given a problem instance (X, f) and a neighborhood function N, a feasible so-lution $x' \in X$ is called locally optimal (minimization problem) with respect to N if

$$f(x') \leq f(x) \quad \text{for all } x \in N(x'). \tag{2.10}$$

Therefore, locally optimal solutions do not exist if no neighborhood is defined. Fur-thermore, the existence of local optima is determined by the neighborhood definition used as different neighborhoods can result in different locally optimal solutions.

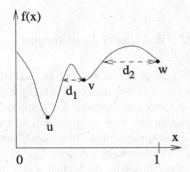

Fig. 2.2 Locally and globally optimal solutions

Figure 2.2 illustrates the differences between locally and globally optimal so-lutions and shows how local optima depend on the definition of N. We have a one-dimensional minimization problem with $x \in [0, 1] \in \mathbb{R}$. We assume an objec-tive function f that assigns objective values to all $x \in X$. Independently of the neighborhood used, u is always the globally optimal solution. If we use the 1-dimensional Euclidean distance (2.6) as metric and define a neighborhood around x as $N(x) = \{y | y \in X \text{ and } d(x, y) \leq \varepsilon\}$ the solution v is a locally optimal solution if $\varepsilon < d_1$. Analogously, w is locally optimal for all neighborhoods with $\varepsilon < d_2$. For $\varepsilon \geq d_2$, the only locally optimal solution is the globally optimal solution u.

The modality of a problem describes the number of local optima in the prob-lem. Unimodal problems have only one local optimum (which is also the global optimum) whereas multi-modal problems have multiple local optima. In general, multi-modal problems are more difficult for guided search methods to solve than unimodal problems.

2.4 Properties of Optimization Problems

The purpose of optimization algorithms is to find high-quality solutions for a prob-lem. If possible, they should identify either optimal solutions x^*, near-optimal solu-tions $x \in X$, where $f(x) - f(x^*)$ is small, or at least locally optimal solutions.

Problem difficulty describes how difficult it is to find an optimal solution for a specific problem or problem instance. Problem difficulty is defined independently of the optimization method used. Determining the difficulty of a problem is often a difficult task as we have to prove that that there are no optimization methods that can better solve the problem. Statements about the difficulty of a problem are method-independent as they must hold for all possible optimization methods.

We know that different types of optimization methods lead to different search performance. Often, optimization methods perform better if they exploit some characteristics of an optimization problem. In contrast, methods that do not exploit any problem characteristics, like black-box optimization techniques, usually show low performance. For an example, we have a look at *random search*. In random search, solutions $x \in X$ are iteratively chosen in a random order. We want to assume that each solution is considered only once by random search. When random search is stopped, it returns the best solution found. During random search, new solutions are chosen randomly and no problem-specific information about the structure of the problem or previous search steps is used. The number of evaluations needed by random search is the number of elements drawn from X, which is independent of the problem itself (if we assume a unique optimal solution). Consequently, a distinction between "easy" and "difficult" problems is meaningless when random search is the only available optimization method.

The following sections study properties of optimization problems. Section 2.4.1 starts with complexity classes which allow us to formulate bounds on the performance of algorithmic methods. This allows us to make statements about the difficulty of a problem. Then, we continue with properties of optimization problems that can be exploited by modern heuristics. Section 2.4.2 introduces the locality of a problem and presents corresponding measurements. The locality of a problem is exploited by guided search methods which perform well if problem locality is high. Important for high locality is a proper definition of a metric on the search space. Finally, Sect. 2.4.3 discusses the decomposability of a problem and how recombination-based optimization methods exploit a problem's decomposability.

2.4.1 Problem Difficulty

The complexity of an algorithm is the effort (usually time or memory) that is necessary to solve a particular problem. The effort depends on the input size, which is equal to the size n of the problem to be solved. The difficulty or complexity of a problem is the lowest possible effort that is necessary to solve the problem.

Therefore, problem difficulty is closely related to the complexity of algorithms. Based on the complexity of algorithms, we are able to find upper and lower bounds on the problem difficulty. If we know that an algorithm can solve a problem, we automatically have an upper bound on the difficulty of the problem, which is just the complexity of the algorithm. For example, we study the problem of finding a friend's telephone number in the telephone book. The most straightforward approach is to search through the whole book starting from "A". The effort for doing this increases linearly with the number of names in the book. Therefore, we have an upper bound

on the difficulty of the problem (problem has at most linear complexity) as we know
a linear algorithm that can solve the problem. A more effective way to solve this
problem is bisection or binary search which iteratively splits the entries of the book
into halves. With n entries, we only need $\log(n)$ search steps to find the number. So,
we have a new, improved, upper bound on problem difficulty.

Finding lower bounds on the problem difficulty is more difficult as we have to
show that no algorithm exists that needs less effort to solve the problem. Our problem of finding a friend's name in a telephone book is equivalent to the problem of
searching an ordered list. Binary search which searches by iteratively splitting the
list into halves is optimal and there is no method with lower effort (Knuth, 1998).
Therefore, we have a lower bound and there is no algorithm that needs less than
$\log(n)$ steps to find an address in a phone book with n entries. A problem is called
closed if the upper and lower bounds on its problem difficulty are identical. Consequently, the problem of searching an ordered list is closed.

This section illustrates how bounds on the difficulty of problems can be derived
by studying the effort of optimization algorithms that are used to solve the problems.
As a result, we are able to classify problems as easy or difficult with respect to the
performance of the best-performing algorithm that can solve the problem.

The following paragraphs give an overview of the Landau notation which is an
instrument for formulating upper and lower bounds on the effort of optimization
algorithms. Thus, we can also use it for describing problem difficulty. Then, we
illustrate that each optimization problem can also be modeled as a decision problem
of the same difficulty. Finally, we illustrate different complexity classes (P, NP, NP-
hard, and NP-complete) and discuss the tractability of decision and optimization
problems.

2.4.1.1 Landau Notation

The Landau notation (which was introduced by Bachmann (1894) and made popular
by the work of Landau (1974)) can be used to compare the asymptotic growth of
functions and is helpful when measuring the complexity of problems or algorithms.
It allows us to formulate asymptotic upper and lower bounds on function values.
For example, Landau notation can be used to determine the minimal amount of
memory or time that is necessary to solve a specific problem. With $n \in \mathbb{N}$, $c \in \mathbb{R}$,
and $f, g : \mathbb{N} \to \mathbb{R}$ the following bounds can be described using the Landau symbols:

- **asymptotic upper bound** ("big O notation"):
  $f \in O(g) \Leftrightarrow \exists c > 0 \; \exists n_0 > 0 \; \forall n \geq n_0 : |f(n)| \leq c|g(n)|$: f is dominated by g.
- **asymptotically negligible** ("little o notation"):
  $f \in o(g) \Leftrightarrow \forall c > 0 \; \exists n_0 > 0 \; \forall n \geq n_0 : |f(n)| < c|g(n)|$: f grows slower than g.
- **asymptotic lower bound**:
  $f \in \Omega(g) \Leftrightarrow g \in O(f)$: f grows at least as fast as g.
- **asymptotically dominant**:
  $f \in \omega(g) \Leftrightarrow g \in o(f)$: f grows faster than g.

- **asymptotically tight bound**:
  $f \in \Theta(g) \Leftrightarrow g \in O(f) \wedge f \in O(g)$: g and f grow at the same rate.

Using this notation, it is easy to compare the difficulty of different problems: using $O(g)$ and $\Omega(g)$, it is possible to give an upper, respectively lower bound for the asymptotic running time f of algorithms that are used to solve a problem. It is important to have in mind that lower bounds on problem difficulty hold for all possible optimization problems for a specific problem. In contrast, upper bounds only indicate that there is at least one algorithm that can solve the problem with this effort but there are also other algorithms where a higher effort is necessary. The Landau notation does not consider constant factors since these mainly depend on the computer or implementation used to solve the problem. Therefore, constants do not directly influence the difficulty of a problem.

We give three small examples. In problem 1, we want to find the smallest number in an unordered list of n numbers. The complexity of this problem is $O(n)$ (it increases linearly with n) when using linear search and examining all possible elements in the list. As it is not possible to solve this problem faster than linear, there is no gap between the lower bound $\Omega(n)$ and upper bound $O(n)$. In problem 2, we want to find an element in an ordered list with n items (for example finding a telephone number in the telephone book). Binary search (bisection) iteratively splits the list into two halves and can find any item in $\log(n)$ search steps. Therefore, the upper bound on the complexity of this problem is $O(\log(n))$. Again, the lower bound is equal to the upper bound (Harel and Rosner, 1992; Cormen et al, 2001) and the complexity of the problem is $\Theta(\log(n))$. Finally, in problem 3 we want to sort an array of n arbitrary elements. Using standard sorting algorithms like merge sort (Cormen et al, 2001) it can be solved in $O(n\log(n))$. As the lower bound is $\Omega(n\log(n))$, the difficulty of this problem is $\Theta(n\log(n))$.

2.4.1.2 Optimization Problems, Evaluation Problems, and Decision Problems

To derive upper and lower bounds on problem difficulty, we need a formal description of the problem. Thus, at least the solutions $x \in X$ and the objective function $f : X \to \mathbb{R}$ must be defined. The search space can be very trivial (e.g. have a trivial topology) as the definition of a neighborhood structure is not necessary. Developing bounds is difficult for problems where the objective function does not systematically assign objective values to each solution. Although describing X and f is sufficient for formulating a problem model, in most problems of practical relevance the size $|X|$ of the search space is large and, thus, the direct assignment of objective values to each possible solution is often very time-consuming and not appropriate for building an optimization model that should be solved by a computer.

To overcome this problem, we can define each optimization problem implicitly using two algorithms $\mathscr{A}_X$ and $\mathscr{A}_f$, and two sets of parameters S_X and S_f. Given a set of parameters S_X, the algorithm $\mathscr{A}_X(x, S_X)$ decides whether the solution x is an element of X, i.e. whether x is a feasible solution. Given a set of parameters S_f, the algorithm $\mathscr{A}_f(x, S_f)$ calculates the objective value of a solution x. Therefore,

$\mathscr{A}_f(x, S_f)$ is equivalent to the objective function $f(x)$. For given $\mathscr{A}_X$ and $\mathscr{A}_f$, we can define different instances of a combinatorial optimization problem by assigning different values to the parameters in S_X and S_f.

We want to give an example. We have a set of s different items and want to find a subset of n items with minimal weight. Then, the algorithm $\mathscr{A}_X$ checks whether a solution x is feasible. The set S_X contains only one parameter which is the number n of items that should be found. If x contains exactly n items, $\mathscr{A}_X(x, S_X)$ indicates that x is a feasible solution. The parameters S_f are the weights w_i of the different items. Algorithm $\mathscr{A}_f$ calculates the objective value of a feasible solution x by summing up the weights of the n items ($f(x, w) = \sum w_i$) using the solution x and the weights w_i. Using these definitions, we can formulate an *optimization problem* as:

> Given are the two algorithms $\mathscr{A}_X$ and $\mathscr{A}_f$ and representations of the parameters S_X and S_f.
> The goal is to find the optimal feasible solution.

This formulation of an optimization problem is equivalent to (2.3) (p. 15). The algorithm $\mathscr{A}_X$ checks whether all constraints are met and $\mathscr{A}_f$ calculates the objective value of x. Analogously, the *evaluation version* of an optimization problem can be defined as:

> Given are the two algorithms $\mathscr{A}_X$ and $\mathscr{A}_f$ and representations of the parameters S_X and S_f.
> The goal is to find the objective value of the optimal solution.

Finally, the *decision version* (also known as recognition version) of an optimization problem can be defined as:

> Given are the algorithms $\mathscr{A}_X$ and $\mathscr{A}_f$, representations of the parameters S_X and S_f, and an integer L. The goal is to decide whether there is a feasible solution $x \in X$ such that $f(x) \leq L$.

The first two versions are problems where an optimal solution has to be found, whereas in the decision version of an optimization problem a question has to be answered either by *yes* or *no*. We denote feasible solutions x whose objective value $f(x) \leq L$ as *yes*-solutions. We can solve the decision version of the optimization problem by solving the original optimization problem, calculating the objective value $f(x^*)$ of the optimal solution x^*, and deciding whether $f(x^*)$ is greater than L. Therefore, the difficulty of the three versions is roughly the same if we assume that $f(x^*)$, respectively $\mathscr{A}_f(x^*, X_f)$, is easy to compute (Papadimitriou and Steiglitz, 1982, Chap. 15.2). If this is the case, all three versions are equivalent.

We may ask why we want to formulate an optimization as a decision problem which is much less intuitive? The reason is that in computational complexity theory, many statements on problem difficulty are formulated for decision (and not optimization) problems. By formulating optimization problems as decision problems, we can apply all these results to optimization problems. Therefore, complexity classes which can be used to categorize decision problems in classes with different difficulty (compare the following paragraphs) can also be used for categorizing optimization problems. For more information on the differences between optimization, evaluation, and decision versions of an optimization problem, we refer the interested reader to Papadimitriou and Steiglitz (1982, Chap. 15) or Harel and Rosner (1992).

2.4.1.3 Complexity Classes

Computational complexity theory ((Hartmanis and Stearns, 1965; Cook, 1971; Garey and Johnson, 1979; Papadimitriou and Yannakakis, 1991; Papadimitriou, 1994; Arora and Barak, 2009) allows us to categorize decision problems in different groups based on their difficulty. The difficulty of a problem is defined with respect to the amount of computational resources that are at least necessary to solve the problem.

In general, the effort (amount of computational resources) that is necessary to solve an optimization problem of size n is determined by its time and space complexity. *Time complexity* describes how many iterations or number of search steps are necessary to solve a problem. Problems are more difficult if more time is necessary. *Space complexity* describes the amount of space (usually memory on a computer) that is necessary to solve a problem. As for time, problem difficulty increases with higher space complexity. Usually, time and space complexity depend on the input size n and we can use the Landau notation to describe upper and lower bounds on them. A *complexity class* is a set of computational problems where the amount of computational resources that are necessary to solve the problem shows the same asymptotic behavior. For all problems that are contained in one complexity class, we can give bounds on the computational complexity (in general, time and space complexity). Usually, the bounds depend on the size n of the problem, which is also called its *input size*. Usually, n is much smaller than the size $|X|$ of the search space. Typical bounds are asymptotic lower or upper bounds on the time that is necessary to solve a particular problem.

Complexity Class P

The complexity class P (P stands for polynomial) is defined as the set of decision problems that can be solved by an algorithm with worst-case polynomial time complexity. The time that is necessary to solve a decision problem in P is asymptotically bounded (for $n > n_0$) by a polynomial function $O(n^k)$. For all problems in P, an algorithm exists that can solve any instance of the problem in time that is $O(n^k)$, for some k. Therefore, all problems in P can be solved effectively in the worst case. As we showed in the previous section that all optimization problems can be formulated as decision problems, the class P can be used to categorize optimization problems.

Complexity Class NP

The class NP (which stands for non-deterministic polynomial time) describes the set of decision problems where a *yes* solution of a problem can be verified in polynomial time. Therefore, both the formal representation of a solution x and the time it takes to check its validity (to check whether it is a *yes* solution) must be polynomial or polynomially-bounded.

Therefore, all problems in NP have the property that their *yes* solutions can be checked effectively. The definition of NP says nothing about the time necessary for verifying *no* solutions and a problem in NP can not necessarily be solved in polynomial time. Informally, the class NP consists of all "reasonable" problems of practical importance where a *yes* solution can be verified in polynomial time: this means the objective value of the optimal solution can be calculated fast. For problems not in NP, even verifying that a solution is valid (is a *yes* answer) can be extremely difficult (needs exponential time).

An alternative definition of NP is based on the notion of *non-deterministic algorithms*. Non-deterministic algorithms are algorithms which have the additional ability to guess any verifiable intermediate result in a single step. If we assume that we find a *yes* solution for a decision problem by iteratively assigning values to the decision variables, a non-deterministic algorithm always selects the value (possibility) that leads to a *yes* answer, if a *yes* answer exists for the problem. Therefore, we can view a non-deterministic algorithm as an algorithm that always guesses the right possibility whenever the correctness can be checked in polynomial time. The class NP is the set of all decision problems that can be solved by a non-deterministic algorithm in worst-case polynomial time. The two definitions of NP are equivalent to each other. Although non-deterministic algorithms cannot be executed directly by conventional computers, this concept is important and helpful for the analysis of the computational complexity of problems.

All problems that are in P also belong to the class NP. Therefore, $P \subseteq NP$. An important question in computational complexity is whether P is a proper subset of NP ($P \subset NP$) or whether NP is equal to P ($P = NP$). So far, this question is not finally answered (Fortnow, 2009) but most researchers assume that $P \neq NP$ and there are problems that are in NP, but not in P.

In addition to the classes P and NP, there are also problems where *yes* solutions cannot be verified in polynomial time. Such problems are very difficult to solve and are, so far, of only little practical relevance.

Tractable and Intractable Problems

When solving an optimization problem, we are interested in the running time of the algorithm that is able to solve the problem. In general, we can distinguish between polynomial running time and exponential running time. Problems that can be solved using a polynomial-time algorithm (there is an upper bound $O(n^k)$ on the running time of the algorithm, where k is constant) are *tractable*. Usually, tractable problems are easy to solve as running time increases relatively slowly with larger input size n. For example, finding the lowest element in an unordered list of size n is tractable as there are algorithms with time complexity that is $O(n)$. Spending twice as much effort solving the problem allows us to solve problems twice as large.

In contrast, problems are *intractable* if they cannot be solved by a polynomial-time algorithm and there is a lower bound on the running time which is $\Omega(k^n)$, where $k > 1$ is a constant and n is the problem size (input size). For example, guessing the correct number for a digital door lock with n digits is an intractable problem, as the time necessary for finding the correct key is $\Omega(10^n)$. Using a lock with one more digit increases the number of required search steps by a factor of 10. For this problem, the size of the problem is n, whereas the size of the search space is $|X| = 10^n$. The effort to find the correct key depends on n and increases at the same rate as the size of the search space. Table 2.2 lists the growth rate of some common functions ordered by how fast they grow.

constant	$O(1)$
logarithmic	$O(\log n)$
Table 2.2 Polynomial (top)	linear	$O(n)$
and exponential (bottom)	quasilinear	$O(n \log n)$
functions	quadratic	$O(n^2)$
polynomial (of order c)	$O(n^c), c > 1$
exponential	$O(k^n)$
factorial	$O(n!)$
super-exponential	$O(n^n)$

We can identify three different types of problems with different difficulty.

- Tractable problems with known polynomial-time algorithms. These are easy problems. All tractable problems are in P.
- Provably intractable problems, where we know that there is no polynomial-time algorithm. These are difficult problems.
- Problems where no polynomial-time algorithm is known but intractability has not yet been shown. These problems are also difficult.

NP-Hard and NP-Complete

All decision problems that are in P are tractable and thus can be easily solved using the "right" algorithm. If we assume that $P \neq NP$, then there are also problems that are in NP but not in P. These problems are difficult as no polynomial-time algorithms exist for them.

Among the decision problems in NP, there are problems where no polynomial algorithm is available and which can be transformed into each other with polynomial effort. Consequently, a problem is denoted *NP-hard* if an algorithm for solving this problem is polynomial-time reducible to an algorithm that is able to solve *any* problem in NP. A problem A is polynomial-time reducible to a different problem B if and only if there is a transformation that transforms any arbitrary solution x of A into a solution x' of B in polynomial time such that x is a *yes* instance for A if and only if x' is a *yes* instance for B. Informally, a problem A is reducible to some other problem B if problem B either has the same difficulty or is harder than problem A. Therefore, NP-hard problems are at least as hard as any other problem in NP, although they might be harder. Therefore, NP-hard problems are not necessarily in NP.

Cook (1971) introduced the set of *NP-complete* problems as a subset of NP. A decision problem A is denoted NP-complete if

- A is in NP and
- A is NP-hard.

Therefore, no other problem in NP is more than a polynomial factor harder than any NP-complete problem. Informally, NP-complete problems are the most difficult problems that are in NP.

All NP-complete problems form one set as all NP-complete problems have the same complexity. However, it is as yet unclear whether NP-complete problems are tractable. If we are able to find a polynomial-time algorithm for any one of the NP-complete problems, then every NP-complete problem can be solved in polynomial time. Then, all other problems in NP can also be solved in polynomial time (are tractable) and thus P = NP. On the other hand, if it can be shown that one NP-complete problem is intractable, then all NP-complete problems are intractable and P ≠ NP.

Summarizing our discussion, we can categorize optimization problems with respect to the computational effort that is necessary for solving them. Problems that are in P are usually easy as algorithms are known that solve such problems in polynomial time. Problems that are NP-complete are difficult as no polynomial-time algorithms are known. Decision problems that are not in NP are even more difficult as we could not evaluate in polynomial time whether a particular solution for such a problem is feasible. To be able to calculate upper and lower bounds on problem complexity, usually well-defined problems are necessary that can be formulated in functional form.

2.4.2 Locality

In general, the *locality* of a problem describes how well the distances $d(x,y)$ between any two solutions $x, y \in X$ correspond to the difference of the objective values $|f(x) - f(y)|$ (Lohmann, 1993; Rechenberg, 1994; Rothlauf, 2006). The locality of a problem is high if neighboring solutions have similar objective values. In contrast, the locality of a problem is low if low distances do not correspond to low differences of the objective values. Relevant determinants for the locality of a problem are the metric defined on the search space and the objective function f.

In the heuristic literature, there are a number of studies on locality for discrete decision variables (Weicker and Weicker, 1998; Rothlauf and Goldberg, 1999, 2000; Gottlieb and Raidl, 2000; Gottlieb et al, 2001; Whitley and Rowe, 2005; Caminiti and Petreschi, 2005; Raidl and Gottlieb, 2005; Paulden and Smith, 2006) as well as for continuous decision variables (Rechenberg, 1994; Igel, 1998; Sendhoff et al, 1997b,a). For continuous decision variables, locality is also known as *causality*. High and low locality correspond to strong and weak causality, respectively. Although causality and locality describe the same concept and causality is the older one, we refer to the concept as locality as it is currently more often used in the literature.

Guided search methods are optimization approaches that iteratively sample solutions and use the objective values of previously sampled solutions to guide the future search process. In contrast to random search which samples solutions randomly and uses no information about previously sampled solutions, guided search methods differentiate between promising (for maximization problems these are solutions with high objective values) and non-promising (solutions with low objective values) areas in the fitness landscape. New solutions are usually generated in the neighborhood of promising solutions with high objective values. A prominent example of guided search is greedy search (see Sect. 3.4.1).

The locality of optimization problems has a strong impact on the performance of guided search methods. Problems with high locality allow guided search methods to find high-quality solutions in the neighborhood of already found good solutions. Furthermore, the underlying idea of guided search methods to move in the search space from low-quality solutions to high-quality solutions works well if the problem has high locality. In contrast, if a problem has low locality, guided search methods cannot make use of previous search steps to extract information that can be used for guiding the search. Then, for problems with low locality, guided search methods behave like random search.

One of the first approaches to the question of what makes problems difficult for guided search methods, was the study of deceptive problems by Goldberg (1987) which was based on the work of Bethke (1981). In deceptive problems, the objective values are assigned in such a way to the solutions that guided search methods are led away from the global optimal solution. Therefore, based on the structure of the fitness landscape (Weinberger, 1990; Manderick et al, 1991; Deb et al, 1997), the correlation between the fitness of solutions can be used to describe how difficult a specific problem is to solve for guided search methods. For an overview of correlation measurements and problem difficulty we refer to Bäck et al (1997, Chap. B2.7) or Reeves and Rowe (2003).

The following paragraphs present approaches that try to determine what makes a problem difficult for guided search. Their general idea is to measure how well the metric defined on the search space fits the structure of the objective function. A high fit between metric and structure of the fitness function makes a problem easy for guided search methods.

2.4.2.1 Fitness-Distance Correlation

A straightforward approach for measuring the difficulty of problems for guided search methods has been presented in Jones and Forrest (1995). They assumed that the difficulty of an optimization problem is determined by how the objective values are assigned to the solutions $x \in X$ and what metric is defined on X. Combining both aspects, problem difficulty can be measured by the *fitness-distance correlation coefficient*

$$\rho_{FDC} = \frac{c_{fd}}{\sigma(f)\sigma(d_{opt})}, \tag{2.11}$$

where

$$c_{fd} = \frac{1}{m} \sum_{i=1}^{m} (f_i - \langle f \rangle)(d_{i,opt} - \langle d_{opt} \rangle)$$

is the covariance of f and d_{opt}. $\langle f \rangle$, $\langle d_{opt} \rangle$, $\sigma(f)$, and $\sigma(d_{opt})$ are the means and standard deviations of the fitness f and the distance d_{opt} to the optimal solution x^*, respectively (Jones, 1995a; Jones and Forrest, 1995; Altenberg, 1997). $d_{i,opt}$ is the distance of solution i to the optimal solution x^*. The fitness-distance correlation coefficient $\rho_{FDC} \in [-1, 1]$ measures the linear correlation between the fitness of search points and their distances to the global optimum x^*.

As ρ_{FDC} represents a summary statistic of f and d_{opt}, it works well if f and d_{opt} follow a bivariate normal distribution. For problems where f and d_{opt} do not follow a normal distribution, using the correlation as a measure of problem difficulty for guided search methods will not yield meaningful results (Jones and Forrest, 1995).

Using the fitness-distance correlation coefficient, Jones and Forrest (1995) classified fitness landscapes (for maximization problems) into three classes, straightforward ($\rho_{FDC} \leq -0.15$), difficult ($-0.15 < \rho_{FDC} < 0.15$), and misleading ($\rho_{FDC} \geq 0.15$):

1. Straightforward: For such problems, the fitness of a solution is correlated with the distance to the optimal solution. With lower distance, the fitness difference to the optimal solution decreases. As the structure of the search space guides search methods towards the optimal solution such problems are usually easy for guided search methods.
2. Difficult: There is no correlation between the fitness difference and the distance to the optimal solution. The fitness values of neighboring solutions are uncorrelated and the structure of the search space provides no information about which solutions should be sampled next by the search method.
3. Misleading: The fitness difference is negatively correlated to the distance to the optimal solution. Therefore, the structure of the search space misleads a guided search method to sub-optimal solutions.

For minimization problems, the situation is reversed as problems are straightforward for $\rho_{FDC} \geq 0.15$, difficult for $-0.15 < \rho_{FDC} < 0.15$, and misleading for $\rho_{FDC} \leq -0.15$. The three different classes of problem difficulty are illustrated in Fig. 2.3. We show how the fitness difference $|f(x^*) - f|$ depends on the distance d_{opt} to the optimal solution x^*. In the following paragraphs, we want to discuss these three classes in some more detail.

Problems are easy for guided search methods if there is a positive correlation between a solution's distance to the optimal solution and the difference between its fitness and the fitness of the optimal solution. An example is a one-dimensional problem where the fitness of a solution is equal to the distance to the optimal solution ($f(x) = d(x, x^*)$). Then, $\rho_{FDC} = 1$ (for a minimization problem) and the problem can easily be solved using guided search methods.

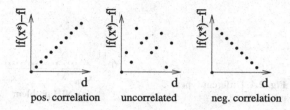

Fig. 2.3 Different classes of problem difficulty

Problems become more difficult if there is no correlation between the fitness difference and the distance to the optimal solution. The locality of such problems is low, as no meaningful relationship exists between the distances d between different solutions and their objective values. Thus, the fitness landscape cannot guide guided search methods to optimal solutions. Optimization methods cannot use information about a problem which was collected in prior search steps to determine the next search step. Therefore, all search algorithms show the same performance as no useful information (information that indicates where the optimal solution can be found) is available for the problem. Because all search strategies are equivalent, also random search is an appropriate search method for such problems. Random search uses no information and performs as well as other search methods on these types of problems.

We want to give two examples. In the first example, we have a discrete search space X with n elements $x \in X$. A deterministic random number generator assigns a random number to each x. Again, the optimization problem is to find x^*, where $f(x^*) \leq f(x)$ for all $x \in X$. Although we can define neighborhoods and similarities between different solutions, all possible optimization algorithms show the same behavior. All elements of the search space must be evaluated to find the globally optimal solution.

The second example is the needle-in-a-haystack (NIH) problem. Following its name, the goal is to find a needle in a haystack. In this problem, a metric exists defining distances between solutions, but there is no meaningful relationship between the metric and the objective value (needle found or not) of different solutions. When physically searching in a haystack for a needle, there is no good strategy for choosing promising areas of the haystack that should be searched in the next search step. The NIH problem can be formalized by assuming a discrete search space X and the objective function

$$f(x) = \begin{cases} 0 & \text{for } x \neq x^{opt} \\ 1 & \text{for } x = x^{opt}. \end{cases} \tag{2.12}$$

Figure 2.4(a) illustrates the problem. The NIH problem is equivalent to the problem of finding the largest number in an unordered list of numbers. The effort to solve such problems is high and increases linearly with the size $|X|$ of the search space. Therefore, the difficulty of the NIH problem is $\Theta(n)$.

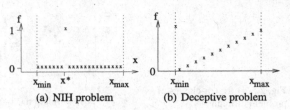

Fig. 2.4 Different types of problems

(a) NIH problem (b) Deceptive problem

Guided search methods perform worst for problems where the fitness landscape leads the search method away from the optimal solution. Then, the distance to the optimal solution is negatively correlated to the fitness difference between a solution and the optimal solution. The locality of such problems is relatively high as most neighboring solutions have similar fitness. However, since guided search finds the optimal solution by performing iterated small steps in the direction of better solutions, all guided search approaches must fail as they are misled. All other search methods that use information about the fitness landscape also fail. More effective search methods for such problems are those that do not use information about the structure of the search space but search randomly, like random search. The most prominent example of such types of problems are deceptive traps (see Figure 2.4(b)). For this problem, the optimal solution is $x^* = x_{min}$. The solution x_{max} is a deceptive attractor and guided search methods that search in the direction of solutions with higher objective function always find x_{max}, which is not the optimal solution.

A common tool for studying the fitness-distance correlation of problems is fitness distance plots. Usually, such plots are more meaningful than just calculating c_{fd}. Fitness distance problems show how the fitness of randomly sampled solutions depends on their distance to the optimal solution. Examples can be found in Kauffman (1989) (NK-landscapes), Boese (1995) (traveling salesman problem), Reeves (1999b) (flow-shop scheduling problems), Inayoshi and Manderick (1994) and Merz and Freisleben (2000b) (graph bipartitioning problem), Merz and Freisleben (2000a) (quadratic assignment problem), or Mendes et al (2002) (single machine scheduling problem).

2.4.2.2 Ruggedness

For studying the fitness-distance correlation of problems, it is necessary to know the optimal solution. However, for real-world problems the optimal solution is not a priori known and other approaches are necessary that describe how well the metric fits the structure of the objective function.

The performance of guided search methods depends on the properties of the fitness landscape like the number of local optima or peaks in the landscape, the distribution of the peaks in the search space, and the height of the different peaks. Consequently, correlation functions have been proposed to measure the ruggedness of a fitness landscape (Kauffman and Levin, 1987; Kauffman, 1989; Weinberger, 1990; Kauffman, 1993). Like in fitness-distance correlation, the idea is to consider the objective values as random variables and to obtain statistical properties on how the distribution of the objective values depends on the distances between solutions. The *autocorrelation function* (which is interchangeable with the autocovariance function if the normalization factor $\langle f^2 \rangle - \langle f \rangle^2$ is dropped) of a fitness landscape is defined as (Merz and Freisleben, 2000b)

$$\rho(d) = \frac{\langle f(x)f(y) \rangle_{d(x,y)=d} - \langle f \rangle^2}{\langle f^2 \rangle - \langle f \rangle^2}, \qquad (2.13)$$

where $\langle f \rangle$ denotes the average value of f over all $x \in X$ and $\langle f(x)f(y) \rangle_{d(x,y)=d}$ is the average value of $f(x)f(y)$ for all pairs $(x,y) \in S \times S$, where $d(x,y) = d$. The autocorrelation function has the attractive property of being in the range $[-1,1]$. An autocorrelation value of 1 indicates perfect correlation (positive correlation) and -1 indicates prefect anti-correlation (negative correlation). For a fixed distance d, ρ is the correlation between the objective values of all solutions that have a distance of d. Weinberger recognized that landscapes with exponentially decaying autocovariance functions are often easy to solve for guided search methods (Weinberger, 1990).

To calculate the autocorrelation function is demanding for optimization problems as it requires evaluating all solutions of the search space. Therefore, Weinberger used random walks through the fitness landscape to approximate the autocorrelation function. A random walk is an iterative procedure where in each search step a random neighboring solution is created. The *random walk correlation function* (Weinberger, 1990; Stadler, 1995, 1996; Reidys and Stadler, 2002) is defined as

$$r(s) = \frac{\langle f(x_i)f(x_{i+s}) \rangle - \langle f \rangle^2}{\langle f^2 \rangle - \langle f \rangle^2}, \qquad (2.14)$$

where x_i is the solution examined in the ith step of the random walk. s is the number of steps between two solutions in the search space. For a fixed s, r defines the correlation of two solutions that are reached by a random walk in s steps, where $s \geq d_{min}$. For a random walk with a large number of steps, $r(s)$ is a good estimate for $\rho(d)$.

Correlation functions have some nice properties and can be used to measure the difficulty of a problem for guided search methods. If we assume that we have a completely random problem, where random objective values are assigned to all $x \in X$, then the autocorrelation function will have a peak at $d = s = 0$ and will be close to zero for all other d and s. In general, for all possible problems the autocorrelation function reaches its peak at the origin $d = s = 0$. Thus it holds $|r(s)| \leq r(0)$ for all $0 < s \leq d_{max}$.

When assuming that the distance between two neighboring solutions x and y is equal to one ($d(x,y) = 1$), $r(1)$ measures the correlation between the objective values of all neighboring solutions. The *correlation length* l_{corr} of a landscape (Stadler, 1992, 1996) is defined as

$$l_{corr} = -\frac{1}{\ln(|r(1)|)} = -\frac{1}{\ln(|\rho(1)|)}$$

for $r(1), \rho(1) \neq 0$.

The ruggedness of a fitness landscape depends on the correlation between neighboring solutions. If the correlation (the correlation length) is high, neighboring solutions have similar objective values and the fitness landscape is smooth and not rugged. For optimization problems where the autocorrelation function indicates strong correlation ($\rho(d) \approx 1$), guided search methods are a good choice as the structure of the search space defined by the metric fits well the structure of the objective function. High-quality solutions are grouped together in the search space and the probability of finding a good solution is higher in the neighborhood of a high-quality solution than in the neighborhood of a low-quality solution.

Correlation functions give us meaningful estimates on how difficult a problem is for guided search only if the search space possesses *regularity*. Regularity means that all elements of the landscape are visited by a random walk with equal probability (Weinberger, 1990). Many optimization problems like the traveling salesman problem (Kirkpatrick and Toulouse, 1985; Stadler and Schnabl, 1992), the quadratic assignment problem (Taillard, 1995; Merz and Freisleben, 2000a), and the flow-shop scheduling problem (Reeves, 1999b) possess regular search spaces. However, other problems like job-shop scheduling problems do not possess this regularity as random walks through the search space are biased (Bierwirth et al, 2004). Such a bias affects random walks and directed stochastic local search algorithms.

In the literature, there are various examples of how the correlation length can be used to study the difficulty of optimization problems for guided search methods (Kauffman and Levin, 1987; Kauffman, 1989; Huynen et al, 1996; Kolarov, 1997; Barnett, 1998; Angel and Zissimopoulos, 1998a, 2000, 1998b, 2001, 2002; Grahl et al, 2007).

2.4.3 Decomposability

The *decomposability* of a problem describes how well the problem can be decomposed into several, smaller subproblems that are independent of each other (Polya, 1945; Holland, 1975; Goldberg, 1989c). The decomposability of a problem is high if the structure of the objective function is such that not all decision variables must be simultaneously considered for calculating the objective function but there are groups of decision variables that can be set independently of each other. It is low if it is not possible to decompose a problem into subproblems with few interdependencies between the groups of variables.

When dealing with decomposable problems, it is important to choose the type and number of decision variables such that they fit the properties of the problem. The fit is high if the variables used result in a problem model where groups of decision variables can be solved independently or, at least, where the interactions between groups of decision variables are low. Given the set of decision variables $D = \{x_1, \ldots, x_l\}$, a problem can be decomposed into several subproblems if the objective value of a solution x is calculated as $f(x) = \sum_{D_s} f(\{x_i | x_i \in D_s\})$, where D_s

are non-intersecting and proper subsets of D ($D_s \subsetneq D, \cup D_s = D$) and $i \in \{1, \ldots, l\}$. Instead of summing the objective values for the subproblems also other functions (e.g. multiplication resulting in $f = \prod_{D_s} f(\{x_i | x_i \in D_s\})$) can be used.

In the previous Sect. 2.4.2, we have studied various measures of locality and discussed how the locality of a problem affects the performance of guided search methods. This section focuses on the decomposability of problems and how it affects the performance of *recombination-based search methods*. Recombination-based search methods solve problems by trying different decompositions of the problem, solving the resulting subproblems, and putting together the obtained solutions for these subproblems to get a solution for the overall problem. High decomposability of a problem usually leads to high performance of recombination-based search methods because solving a larger number of smaller subproblems is usually easier than solving the larger, original problem. Consequently, it is important for effective recombination-based search methods to identify proper subsets of variables such that there are no strong interactions between the variables of the different subsets.

We discussed in Sect. 2.3.2 that the type and number of decision variables influences the resulting neighborhood structure. In the case of recombination-based search methods, we must choose the decision variables such that the problem can be easily decomposed by the search method. We want to give an example of how the decomposition of a problem can make a problem easier for recombination-based search methods. Imagine you have to design the color and material of a chair. For each of the two design variables, there are three different options. The quality of a design is evaluated by marketing experts that assign an objective value to each combination of color and material. Overall, there are $3 \times 3 = 9$ possible chair designs. If the problem cannot be decomposed, the experts have to evaluate all nine different solutions to find the optimal design. If we assume that the color and material are independent of each other, we (or a recombination-based search method) can try to decompose the problem and separately solve the decomposed subproblems. If the experts separately decide about color and material, the problem becomes easier as only $3 + 3 = 6$ designs have to be evaluated.

Therefore, the use of recombination-based optimization methods suggests that we should define the decision variables of a problem model such that they allow a decomposition of the problem. The variables should be chosen such that there are no (or at least few) interdependencies between different sets of variables. We can study the importance of choosing proper decision variables for the chair example. The first variant assumes no decomposition of the problem. We define one decision variable $x \in X$, where $|X| = 9$. There are nine different solutions and non-recombining optimization methods have to evaluate all possible solutions to find the optimal one. In the second variant, we know that the objective function of the problem can be decomposed. Therefore, we choose two decision variables $x_1 \in X_1 = \{y, b, g\}$ (yellow, blue, and green) and $x_2 \in X_2 = \{w, m, p\}$ (wooden, metal, or plastics), where $|X_1| = |X_2| = 3$. A possible decomposition for the example problem is $f = f_1(x_1) + f_2(x_2)$ (see Table 2.3). Decomposing the problem in such a way results in two subproblems f_1 and f_2 of size $|X_1| = |X_2| = 3$. Comparing the two problem formulations shows that the resulting objective values f of different solu-

tions are the same for both formulations. However, the size of the resulting search
space is lower for the second variant. Therefore, the problem becomes easier to
solve for recombination-based search methods as the assumed problem decomposition ($f = f_1 + f_2$) fits well the properties of the problem.

without problem decomposition $f = f(x_1, x_2)$	additive problem decomposition $f = f_1(x_1) + f_2(x_2)$
$f(y,w) = 3, f(y,m) = 2, f(y,p) = 1,$	$f_1(y) = 0, f_2(w) = 3,$
$f(b,w) = 4, f(b,m) = 3, f(b,p) = 2,$	$f_1(b) = 1, f_2(m) = 2,$
$f(g,w) = 5, f(g,m) = 4, f(g,p) = 3.$	$f_1(g) = 2, f_2(p) = 1.$

Table 2.3 Two different problem formulations

We want to give another example and study two different problems with l binary
decision variables $x_i \in \{0,1\}$ ($|X| = 2^l$). In the first problem, a random objective
value is assigned to each $x \in X$. This problem cannot be decomposed. In the second problem, the objective value of a solution is calculated as $f = \sum_{i=1}^{l} x_i$. This
example problem can be decomposed. Using recombination-based search methods
for the first example is not helpful as no decomposition of the problem is possible. Therefore, all efforts of recombination-based search methods to find proper
decompositions of the problem are useless. The situation is different for the second
example. Recombination-based methods should be able to correctly decompose the
problem and to solve the l subproblems. If the decomposition is done properly by
the recombination-based search method, only $2l$ different solutions need to be evaluated and the problem becomes much easier to solve once the correct decomposition
of the problem is found. However, usually there is additional effort necessary for
finding the correct decomposition.

We see that the choice of proper decision variables is important for the decomposability of an optimization problem. In principle, there are two extremes for combinatorial optimization problems. The one extreme is to encode all possible solu-
tions $x \in X$ using only one decision variable x_1, where $x_1 \in \{1, \ldots, |X|\}$. Using such
a problem model, no decomposition is possible as only one decision variable ex-
ists. At the other extreme, we could use $\log_2 |X|$ binary decision variables encoding
the $|X|$ different solutions. Then, the number of possible decompositions becomes
maximal (there are $2^{\log |X|}$ possible decompositions of the problem). Proper decision
variables for an optimization model should be chosen such that they allow a high
decomposition of the problem. Problem decomposition is problem-specific and depends on the properties of f. We should have in mind that using a different number
of decision variables not only influences problem decomposition but also results in
a different neighborhood (see also Sect. 2.3.2).

In the following paragraphs, we discuss different approaches developed in the
literature to estimate how well a problem can be solved using recombination-based
search methods. All approaches assume that search performance is higher if a problem can be decomposed into smaller subproblems. Section 2.4.3.1 presents polynomial problem decomposition and Sect. 2.4.3.2 illustrates the Walsh decomposition

of a problem. Finally, Sect. 2.4.3.3 discusses schemata and building blocks and how they affect the performance of recombination-based search methods.

2.4.3.1 Polynomial Decomposition

The *linearity of an optimization problem* (which is also known as *epistasis*) can be measured by its polynomial decomposition. Epistasis is low if the linear separability of a problem is high. Epistasis measures the interference between decision variables and describes how well a problem can be decomposed into smaller sub-problems (Holland, 1975; Davidor, 1989, 1991; Naudts et al, 1997). For binary decision variables, any objective function f defined on l decision variables $x_i \in \{0,1\}$ can be decomposed into

$$f(x) = \sum_{N \subset \{1,...,l\}} \alpha_N \prod_{n \in N} \mathbf{e}_n^T x,$$

where the vector $\mathbf{e}_n$ contains 1 in the nth column and 0 elsewhere, T denotes transpose, and the α_N are the coefficients (Liepins and Vose, 1991). Regarding $x = (x_1, \ldots, x_l)$, we may view f as a polynomial in the variables $x_1, \ldots, x_l$. The coefficients α_N describe the non-linearity of the problem. If there are high order coefficients, the problem function is non-linear. If a decomposed problem has only order 1 coefficients, then the problem is linear decomposable. It is possible to determine the maximum non-linearity of $f(x)$ by its highest polynomial coefficients. The higher the order of the α_N, the more non-linear the problem is.

There is some correlation between the non-linearity of a problem and its difficulty for recombination-based search methods (Mason, 1995). However, as illustrated in the following example, there could be high order α_N although the problem can still easily be solved by recombination-based search methods. The function

$$f(x) = \begin{cases} 1 & \text{for } x_1 = x_2 = 0, \\ 2 & \text{for } x_1 = 0, x_2 = 1, \\ 4 & \text{for } x_1 = 1, x_2 = 0, \\ 10 & \text{for } x_1 = x_2 = 1, \end{cases} \tag{2.15}$$

can be decomposed into $f(x) = \alpha_1 + \alpha_2 x_1 + \alpha_3 x_2 + \alpha_4 x_1 x_2 = 1 + 3x_1 + x_2 + 5x_1 x_2$. The problem is decomposable and, thus, easy for recombination-based search methods as each of the two decision variables can be solved independently of each other. However, as the problem is non-linear and high order coefficients exist, the polynomial decomposition wrongly classifies the problem as difficult. This misclassification is due to the fact that the polynomial decomposition assumes a linear decomposition and cannot appropriately describe non-linear dependencies.

2.4.3.2 Walsh Decomposition

Instead of decomposing an objective function into its polynomial coefficients, binary optimization problems can also be decomposed into the corresponding Walsh coefficients. The Walsh transformation is analogous to the discrete Fourier transformation but for functions with binary decision variables. Every real-valued function $f : \{0,1\}^l \to \mathbb{R}$ over l binary decision variables x_i can be expressed as:

$$f(x) = \sum_{j=0}^{2^l-1} w_j\, \psi_j(x).$$

The *Walsh functions* $\psi_j : \{0,1\}^l \to \{-1,1\}$ form a set of 2^l orthogonal functions. The weights $w_j \in \mathbb{R}$ are called *Walsh coefficients*. The indices j are binary strings of length l representing the integers ranging from 0 to $2^l - 1$. The jth Walsh function is defined as:

$$\psi_j(x) = (-1)^{bc(j \wedge x)},$$

with x, j are binary strings and elements of $\{0,1\}^l$, $\wedge$ denotes the bitwise logical AND, and $bc(x)$ is the number of 1 bits in x (Goldberg, 1989a,b; Vose and Wright, 1998a,b). The Walsh coefficients can be computed by the Walsh transformation:

$$w_j = \frac{1}{2^l} \sum_{k=0}^{2^l-1} f(k)\psi_j(k),$$

where k is a binary string of length l. The coefficients w_j measure the contribution to the objective function by the interaction of the binary decision variables x_i indicated by the positions of the 1's in j. With increasing number of 1's in the binary string j, we have more interactions between the binary decision variables x_i. For example, w_{100} and w_{010} measure the linear contribution to f associated with the decision variable x_0 and x_1, respectively. Analogously, w_{001} is the linear contribution of decision variable x_2. w_{111} measures the nonlinear interaction between all three decision variables x_0, x_1, and x_2. Any function f over a discrete $\{0,1\}^l$ search space can be represented as a weighted sum of all possible 2^l Walsh functions ψ_j.

Walsh coefficients are sometimes used to estimate the expected performance of recombination-based search algorithms (Goldberg, 1989a; Oei, 1992; Goldberg, 1992; Reeves and Wright, 1994; Heckendorn et al, 1996). It is known that problems are easy for recombination-based search algorithms like genetic algorithms (see Sect. 5.2.1, p. 147) if a problem has only Walsh coefficients of order 1. Furthermore, difficult problems tend to have higher order Walsh coefficients. However, analogously to the linear polynomial decomposition, the highest order of a coefficient w_i does not allow us an accurate prediction of problem difficulty. This behavior is expected as Walsh functions are polynomials (Goldberg, 1989a,b; Liepins and Vose, 1991).

The insufficient measurement of problem difficulty for recombination-based search methods can be illustrated by the example (2.15). The Walsh coefficients are $w = (4.25, -1.75, -2.75, 1.25)$. Although the problem is easy to solve for recombination-based search methods (x_1 and x_2 can be set independently of each other), there are high-order Walsh coefficients ($w_{11} = 1.25$) which indicate high problem difficulty.

Walsh analysis not only overestimates problem difficulty but also underestimates it. For example, MAX-SAT problems (see Sect. 4.4, p. 126) are difficult (APX-hard, see Sect. 3.4.2), but have only low-order Walsh coefficients (Rana et al, 1998). For example, the MAX-3SAT problem has no coefficients of higher order than 3 and the number of non-zero coefficients of order 3 is low (Rana et al, 1998). Although, Walsh coefficients indicate that the problem is easy, recombination-based search methods cannot perform well for this difficult optimization problem (Rana et al, 1998; Rana and Whitley, 1998; Heckendorn et al, 1996, 1999).

2.4.3.3 Schemata Analysis and Building Blocks

Schemata analysis is an approach developed and commonly used for measuring the difficulty of problems with respect to genetic algorithms (GA, Sect. 5.2.1). As the main search operator of GAs is recombination, GAs are a representative example of recombination-based search methods. Schemata are usually defined for binary search spaces and thus schemata analysis is mainly useful for problems with binary decision variables. However, the idea of building blocks is also applicable to other search spaces (Goldberg, 2002). In the following paragraphs, we introduce schemata and building blocks and describe how these concepts can be used for estimating the difficulty of problems for recombination-based search methods.

Schemata

Schemata were first proposed by Holland (1975) to model the ability of GAs to process similarities between binary decision variables. When using l binary decision variables $x_i \in \{0, 1\}$, a schema $H = [h_1, h_2, \ldots, h_l]$ is a sequence of symbols of length l, where $h_i \in \{0, 1, *\}$. $*$ denotes the "don't care" symbol and tells us that a decision variable is not fixed. A schema stands for the set of solutions which match the schema at all the defined positions, i.e., those positions having either a 0 or a 1. Schemata of this form allow for coarse graining (Stephens and Waelbroeck, 1999; Contreras et al, 2003), where whole sets of strings can be treated as a single entity.

A position in a schema is fixed if there is either a 0 or a 1 at this position. The size or order $o(H)$ of a schema H is defined as the number of fixed positions (0's or 1's) in the schema string. The defining length $\delta(H)$ of a schema H is defined as the distance between (meaning the number of bits that are between) the two outermost fixed bits. The fitness $f_s(H)$ of a schema is defined as the average fitness of all instances of this schema and can be calculated as

$$f_s(H) = \frac{1}{||H||} \sum_{\mathbf{x} \in H} f(\mathbf{x}),$$

where $||H||$ is the number of solutions $\mathbf{x} \in \{0,1\}^l$ that are instances of the schema H. The instances of a schema H are all solutions $\mathbf{x} \in H$. For example, $\mathbf{x} = (0,1,1,0,1)$ and $\mathbf{y} = (0,1,1,0,0)$ are instances of $H = [0*1**]$. The number of solutions that are instances of a schema H can be calculated as $2^{l-o(H)}$. For a more detailed discussion of schemata in the context of GA, we refer to Holland (1975), Goldberg (1989c), Altenberg (1994) or Radcliffe (1997).

Building Blocks

Based on schemata, Goldberg (1989c, p. 20 and p. 41) defined *building blocks* (BB) as "highly fit, short-defining-length schemata". Although BBs are commonly used (especially in the GA literature) they are rarely defined. We can describe a BB as a solution to a subproblem that can be expressed as a schema. Such a schema has high fitness and its size is smaller than the length l of the binary solution. By combining BBs of lower order, recombination-based search methods like GAs can form high-quality over-all solutions.

We can interpret BBs also from a biological perspective and view them as genes. A gene consists of one or more alleles and can be described as a schema with high fitness. Often, genes do not strongly interact with each other and determine specific properties of individuals like hair or eye color.

BBs can be helpful for estimating the performance of recombination-based search algorithms. If the sub-solutions to a problem (the BBs) are short (low $\delta(H)$) and of low order (low $o(H)$), then the problem is assumed to be easy for recombination-based search.

BB-Based Problem Difficulty

Goldberg (2002) presented an approach for problem difficulty based on schemata and BBs. He decomposed problem difficulty for recombination-based search methods like genetic algorithms into

- difficulty within a building block (intra-BB difficulty),
- difficulty between building blocks (inter-BB difficulty), and
- difficulty outside of building blocks (extra-BB difficulty).

This decomposition of problem difficulty assumes that difficult problems are challenging for methods based on building blocks. In the following paragraphs, we briefly discuss these three aspects of BB-difficulty.

If we count the number of schemata of order $o(H) = k$ that have the same fixed positions, there are 2^k different schemata. Viewing a BB of size k as a subproblem, there are 2^k different solutions to this subproblem. Such subproblems cannot be decomposed any more and usually guided or random search methods are applied to find the correct solution for the decomposed subproblems.

Therefore, *intra-BB difficulty* depends on the locality of the subproblem. As discussed in Sect. 2.4.2, (sub)problems are most difficult to solve if the structure of the fitness landscape leads guided search methods away from the optimal solution. Consequently, the deceptiveness (Goldberg, 1987) of a subproblem (for an example of a deceptive problem, see Fig. 2.4(b), p. 34) is at the core of intra-BB difficulty. We can define the deceptiveness of a problem not only by the correlation between objective function and distance (as we have done in Sect. 2.4.2) but also by using the notion of BBs. A problem is said to be deceptive of order k_{max} if for $k < k_{max}$ all schemata that contain parts of the best solution have lower fitness than their competitors (Deb and Goldberg, 1994). Schemata are competitors if they have the same fixed positions. An example of four competing schemata of size $k = 2$ for a binary problem of length $l = 4$ are $H_1 = [0*0*]$, $H_2 = [0*1*]$, $H_3 = [1*0*]$, and $H_4 = [1*1*]$. Therefore, the highest order k_{max} of the schemata that are not misleading determines the intra-BB difficulty of a problem. The higher the maximum order k_{max} of the schemata, the higher is the intra-BB difficulty.

Table 2.4 Average schema fitness for example described by (2.15)

order	2	1		0
schema	11	1*	*1	**
fitness	**10**	**7**	**6**	4.25
schema	01 10 00	0*	*0	
fitness	2 4 1	1.5	2.5	

Table 2.4 shows the average fitness of the schemata for the example from (2.15). All schemata that contain a part of the optimal solution are above average (printed bold) and better than their competitors. Calculating the deceptiveness of the problem based on the fitness of the schemata correctly classifies this problem as very easy.

When using this concept of BB-difficulty for estimating the difficulty of a problem for recombination-based search methods, the most natural and direct way to measure problem difficulty is to analyze the size and length of the BB in the problem. The intra-BB difficulty of a problem can be measured by the maximum length $\delta(H)$ and size $k = o(H)$ of its BBs H (Goldberg, 1989c). Representative examples of use of these concepts to estimate problem difficulty can be found in Goldberg (1992), Radcliffe (1993), or Horn (1995).

Recombination-based search methods solve problems by trying different problem decompositions and solving the resulting subproblems. If a problem is correctly decomposed, optimal solutions (BBs) of the subproblems can be determined independently of each other. Often, the contributions of different subproblems to the overall objective value of a solution is non-uniform. Non-uniform contributions of subproblems to the objective value of a solution determine *inter-BB difficulty*. Problems become more difficult if some BBs have a lower contribution to the objective value of a solution. Furthermore, problems often cannot be decomposed into completely separated and independent sub-problems, but have some interdependencies between subproblems which are an additional source of inter-BB difficulty.

Sources of *extra-BB difficulty* for recombination-based search methods are factors like noise. Non-deterministic noise can randomly modify the objective values of solutions and make the problem more difficult for recombination-based search methods as no accurate decisions can be made on the optimal solutions for the different subproblems. A similar problem occurs if the evaluation of the solutions is non-stationary. Non-stationary environments result in solutions that have different evaluation values at different moments in time.

Chapter 3
Optimization Methods

The previous chapter introduced relevant properties of optimization problems. In this chapter, we describe how optimization problems can be solved and which different types of optimization methods exist for discrete optimization problems. The goal of optimization methods is to find an optimal or near-optimal solution with low computational effort. The effort of an optimization method can be measured as the time (computation time) and space (computer memory) that is consumed by the method. For many optimization methods, and especially for modern heuristics, there is a trade-off between solution quality and effort, as with increasing effort solution quality increases.

We can distinguish between two different types of optimization methods: Exact optimization methods that guarantee finding an optimal solution and heuristic optimization methods where we have no guarantee that an optimal solution is found. Usually, an exact optimization method is the method of choice if it can solve an optimization problem with effort that grows polynomially with the problem size. We have seen in Sect. 2.4.1 that such problems belong to the class P. The situation is different if problems are NP-hard as then exact optimization methods need exponential effort. Then, even medium-sized problem instances often become intractable and cannot be solved any more using exact methods. To overcome these problems, we can use heuristic optimization methods. Usually, such optimization methods are problem-specific as they exploit properties of the problem. Furthermore, they often show good performance for many NP-complete problems and problems of practical relevance.

This chapter starts with analytical and numerical optimization methods. Analytical methods can only be applied to highly structured problems, where the objective function is explicitly known. In Sect. 3.2, we continue with optimization methods for linear, continuous optimization problems. Although our focus is on optimization methods for discrete problems, we need some knowledge about continuous methods as many combinatorial optimization methods are extensions and modifications of continuous optimization methods. We provide an overview and describe the functionality of the Simplex method and interior point methods. Both approaches can be used for solving linear optimization problems, where the objective function and the constraints are continuous and linear. As interior point methods can solve such problems with polynomial effort, such problems belong to the class P. Section 3.3 gives an overview of exact optimization methods for discrete optimization problems. Many linear, discrete optimization problems are NP-complete. We discuss and illustrate the functionality of enumeration methods like informed search methods and branch and bound, dynamic programming methods, and cutting plane methods. The large drawback of such exact methods is that their effort to solve NP-complete problems increases exponentially with the problem size. Finally, Sect. 3.4 gives an overview of heuristic optimization methods. Important types of heuristic optimization methods are approximation algorithms and modern heuristics. Approximation algorithms are heuristics where we have a bound on the quality of the returned solution. Modern heuristics are general-purpose heuristics that use sophisticated intensification and diversification mechanisms for searching through the search space. The relevant elements of modern heuristics are the representation and variation operators used, the fitness function, the initial solution, and the search strategy. These elements are discussed in the subsequent chapters. Finally, we discuss the no-free-lunch theorem which states that general-purpose problem solving is not possible but heuristic optimization methods must exploit specific properties of a problem to outperform random search.

3.1 Analytical and Numerical Optimization Methods

Finding optimal solutions for optimization problems can be relatively easy if the problem is well-defined and there are no constraints on the decision variables. A common example is continuous optimization problems

$$\text{minimize} \quad f(x) : \mathbb{R}^n \to \mathbb{R},$$

where $x \in \mathbb{R}^n$ and $f(x)$ is a twice-differentiable objective function assigning an objective value to each solution. We assume that there are no additional constraints on the decision variables.

Such problems can be solved by finding solutions where the gradient (denoted as ∇ or grad) of the objective function is zero. The gradient of $f(x)$ is a column vector of length n whose columns are the partial derivatives of $f(x)$:

$$\nabla f(x) = \left(\frac{\partial f}{\partial x_1}, \ldots, \frac{\partial f}{\partial x_n} \right)^T$$

For all local maxima and minima of the problem, $\nabla f(x) = 0$. All solutions where $\nabla f(x) = 0$ are called *stationary points*.

There are several options available to determine whether a stationary point is a local minimum or maximum. A simple and straightforward approach is to calculate the objective values of solutions that are in each dimension slightly (infinitesimally) larger and smaller than the stationary point. If the objective values of all solutions are larger than the objective value of the stationary point, then we have a minimum; if they are all smaller, we have a maximum. If some of them are smaller and some are larger, we have a *saddle point*. A saddle point is a stationary point that is no extremum. Figure 3.1 illustrates the saddle point $(0,0)$ and the minimum $(1,-1)$ for the objective function $f(x) = 3x^4 - 4x^3$.

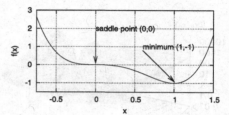

Fig. 3.1 Saddle point and minimum for $f(x) = 3x^4 - 4x^3$

An equivalent approach is calculating the *Hessian matrix* of the objective function $f(x)$ as

$$H(f) = \begin{bmatrix} \frac{\partial^2 f}{\partial x_1^2} & \frac{\partial^2 f}{\partial x_1 \partial x_2} & \cdots & \frac{\partial^2 f}{\partial x_1 \partial x_n} \\ \frac{\partial^2 f}{\partial x_2 \partial x_1} & \frac{\partial^2 f}{\partial x_2^2} & \cdots & \frac{\partial^2 f}{\partial x_2 \partial x_n} \\ \vdots & \vdots & \ddots & \vdots \\ \frac{\partial^2 f}{\partial x_n \partial x_1} & \frac{\partial^2 f}{\partial x_n \partial x_2} & \cdots & \frac{\partial^2 f}{\partial x_n^2} \end{bmatrix}$$

If the determinant of the Hessian matrix is unequal to zero and the Hessian matrix is positive definite at a stationary point y, then y is a local minimum. A matrix is *positive definite* at y if all eigenvalues of the matrix are larger than zero. Analogously, a stationary point is a local maximum if the Hessian matrix is negative definite at y. If the Hessian matrix is indefinite at y (some eigenvalues are positive and some are negative), y is a saddle point. After determining all local minima and maxima of a problem, the optimal solution can be found by comparing the objective values of all local optima and choosing the solution with minimal or maximal objective value, respectively.

If we assume that there are some additional constraints on the range of the decision variables ($x_i \in [x_{min}, x_{max}]$), we can apply the same procedure. In addition, we have to examine all solutions at the edge of the feasible search space ($x_i = x_{min}$ and $x_i = x_{max}$) and check whether they are the global optimum. In general, all solutions at the boundary of the definition space are treated as stationary points.

We want to give an example. *Rosenbrock's function* is a well known test problem for continuous optimization methods (De Jong, 1975; Goldberg, 1989c). It is twice-differentiable and defined as

$$f(x) = \sum_{i=1}^{n-1} [100(x_i^2 - x_{i+1})^2 + (x_i - 1)^2]. \tag{3.1}$$

The global minimum of this problem is $x^* = (1,\ldots,1)$ with the function value $f(x^*) = 0$. For $n < 4$, there are no other local optima. For higher dimensions ($n \geq 4$), the function becomes multi-modal and local optima exist (Hansen and Ostermeier, 2001; Deb et al, 2002; Shang and Qiu, 2006). The problem is viewed as difficult to solve for local search approaches as the minimum resides in a long, narrow, and parabolic-shaped flat valley. Figure 3.2 plots the two-dimensional variant $f(x_1,x_2) = 100(x_1^2 - x_2)^2 + (x_1 - 1)^2$ of Rosenbrock's function for $x_1, x_2 \in [-2,2]$.

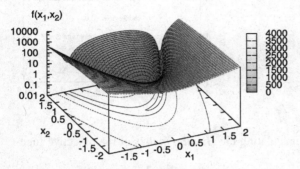

Fig. 3.2 2-dimensional Rosenbrock function

We can calculate the gradient of the n-dimensional Rosenbrock function as

$$\nabla f(x) = \begin{pmatrix} 400x_1(x_1^2 - x_2) + 2(x_1 - 1) \\ \vdots \\ -200(x_{i-1}^2 - x_i) + 400x_i(x_i^2 - x_{i+1}) + 2(x_i - 1) \\ \vdots \\ -200(x_{n-1}^2 - x_n) \end{pmatrix}.$$

For $\frac{\partial f(x)}{\partial x_i} = 0$ ($i = \{1,\ldots,n\}$), we get a system of equations. Solving these equations is difficult for $n \geq 4$, and mainly numerical methods are used (Shang and Qiu, 2006). The resulting stationary point $x^* = (1,\ldots,1)$ is a local (and also global) minimum, since the Hessian matrix is positive definite for $x = x^*$ and the determinant of the Hessian matrix is unequal to zero. For the two-dimensional case, we can calculate x^* analytically. We get

$$\nabla f(x_1,x_2) = \left(400x_1(x_1^2 - x_2) + 2(x_1 - 1), -200(x_1^2 - x_2)\right)^T$$

For $\frac{\partial f}{\partial x_1} = \frac{\partial f}{\partial x_2} = 0$, we get

$$400x_1(x_1^2 - x_2) + 2(x_1 - 1) = 0$$

and

$$-200(x_1^2 - x_2) = 0.$$

The second equation yields $x_2 = x_1^2$. Substituting this into the first equation, we finally get $x_1 = x_2 = 1$, which is the globally optimal solution.

The example illustrates how optimal solutions can be found for problems where the objective function is twice-differentiable. We have to identify stationary points where the gradient of $f(x)$ is zero, exclude saddle points from consideration, and find the optimal solutions among all local optima. However, for many problems we are not able to solve the equation system $\nabla f(x) = 0$ although the objective function is available in functional form and twice-differentiable.

For solving such problems, we can use numerical methods to identify stationary points. Numerical methods often start with an initial solution x^0 and perform iterative steps in the direction of a local optimal solution using the gradient $\nabla f(x)$ for guiding the search. Common examples are *Newton's method* (also known as Newton-Raphson method) and *gradient search*. Newton's method generates new solutions as

$$x^{n+1} = x^n - \gamma [H(f(x^n))]^{-1} \nabla f(x^n), \tag{3.2}$$

where $n \geq 0$, $[H(f(x^n))]^{-1}$ is the inverse of the Hessian matrix at x^n, and γ is a small step size ($\gamma > 0$). For $n \to \infty$, this sequence converges towards a local optimum. Gradient search also performs iterative steps and generates new solutions as

$$x^{n+1} = x^n - \gamma_n \nabla f(x^n), \tag{3.3}$$

where the variable step size γ_n is reduced with larger n such that (minimizing search) $f(x^{n+1}) < f(x^n)$. Again, this sequence converges to a locally optimal solution for $n \to \infty$.

Both numerical methods allow us to find a local optimum in the neighborhood of x^0. Usually, we have to guess an area where the optimal solution can be found ($d(x^*, x^0)$ should be small) and the methods return the nearest local optimum. If we are lucky, the local optimum is also the global optimum. However, as the methods just "move downhill" towards the next local optimum, choosing a random initial solution only returns the optimal solution x^* for unimodal problems. For multi-modal problems, we have no guarantee that the local optimum found is the global optimum.

Summarizing, for some well-defined and twice-differentiable optimization problems we can analytically determine the optimal solution. However, the range of problems that can be solved in such a way is limited. We can extend the number of solvable problems by using numerical methods that use the gradient of the objective function to direct their search through the search space. However, the use of numerical methods is also limited as the effort for searching through the search space and identifying all local optima is high.

3.2 Optimization Methods for Linear, Continuous Problems

Since many problems of practical relevance can be modeled as linear optimization problems, linear optimization methods are an important research theme in OR and are well established in many companies for planning and optimization purposes. Problems are *linear* if

- the objective function depends linearly on the decision variables and
- all relations among the variables are linear.

In linear problems, we assume a number of limited resources (each decision variable describes the consumption of one of the resources) and an objective function that models how the quality of the overall solution depends on the use of the limited resources. Representative examples of linear problems are resource allocation problems, production problems, or network flow problems.

Linear problems are commonly called *linear programming* (LP) problems. The word "programming" is not used in the sense of computer programming (where programming usually means writing a set of computer instructions) but in the sense of "planning". This is due to the fact that "linear programming" was introduced very early and before the word "programming" was associated with computer software.

Optimization methods for linear problems have a very long tradition and go back to early Chinese work from the first century BC (Kangshen et al, 2000). Maclaurin was the first European who presented an approach for solving a system of linear equations which is today known as *Cramer's rule* (Maclaurin, 1748; Cramer, 1750). Around 1800, Gauss reinvented a method already described by Liu Hui in 263 AD (Kangshen et al, 2000) which allows determination of solutions for a system of linear equations and which is today known as *Gaussian elimination*. It took until the emergence of computers (around 1940) to be able to efficiently solve a larger number of linear equations. In 1939, Kantorovich formulated and solved an LP problem (Kantorovich, 1939). In 1947, the US Air Force initiated a project named SCOOP (Scientific Computing of Optimum Programs) in which mathematical techniques were applied to military budgeting and planning. As part of this work, Dantzig proposed a mathematical model of the general linear programming problem and a solution method called the *Simplex method* (Dantzig, 1949, 1951, 1962). In Sect. 3.2.2, we briefly describes the core ideas behind this approach. In 1968, Fiacco and McCormick introduced *interior point methods* (see also Sect. 3.2.3). Such methods could also be used for solving LPs (Khachian, 1979; Karmarkar, 1984) and, in contrast to the Simplex method, their time complexity is polynomial. Today, a thorough understanding of LP problems and methods exists and such methods are widely used in the real world. Optimization methods for LPs are well established and users can choose from a variety of available LP solvers (Mittelmann, 2010).

This section describes linear optimization problems and gives a brief overview of concepts that can be used for solving such problems. Section 3.2.1 formalizes the concept of linear problems, Sect. 3.2.2 describes the basic concept of the Simplex method, and Sect. 3.2.3 compares interior point methods to the Simplex method.

3.2.1 Linear Optimization Problems

In linear, continuous problems, we want to find an optimal solution for a problem, where the objective function depends linearly on the continuous decision variables $x = (x_1, \ldots, x_n) \in \mathbb{R}^n$. A *linear function* is defined as

$$f(x) = c_1 x_1 + c_2 x_2 + \ldots + c_n x_n = \sum_{j=1}^{n} c_j x_j.$$

In addition, there can be linear constraints. We can distinguish between linear equality constraints $f_i(x) = b_i$, where f_i are linear functions, and linear inequality constraints $f_i(x) \geq b_i$ and $f_i(x) \leq b_i$, respectively. The sets of linear equalities and inequalities can be formulated in matrix notation as $Ax = b$, $Ax \leq b$, or $Ax \geq b$, where b is an n-dimensional vector (usually $b \geq 0$) and A is an $n \times m$ matrix. m denotes the number of constraints. The combined linear equalities and inequalities are called *linear constraints*.

In LPs, different problem formulations exist which are equivalent to each other. The *canonical form* of an LP consists of a linear objective function, linear inequalities, and non-negative decision variables:

$$\min \quad c^T x, \tag{3.4}$$
$$\text{subject to} \quad Ax \geq b,$$
$$x_i \geq 0.$$

By multiplying the constraints and the objective function by -1 a linear minimization problem can also be formulated as a maximization problem:

$$\max \quad c^T x, \tag{3.5}$$
$$\text{subject to} \quad Ax \leq b,$$
$$x_i \geq 0$$

In both cases, we assume that the vector x of decision variables is non-negative. Linear optimization problems in canonical form can be transformed into *standard form* (in the context of LPs also known as augmented or slack form):

$$\max \quad c^T x, \tag{3.6}$$
$$\text{subject to} \quad Ax = b,$$
$$x_i \geq 0.$$

To transform a problem from canonical form to standard form, *slack* and *surplus variables* are used. With the slack variables y_i and the surplus variables z_i ($y_i, z_i \geq 0$), we can rewrite inequalities of the form $f(x) \leq b_i$ or $f(x) \geq b_i$ as equalities $f(x) + y_i = b_i$ or $f(x) - z_i = b_i$, respectively. Consequently, each problem in canonical form can be transformed into a problem in standard form with the same properties.

When formulating an LP model of a problem, we have to ensure that

- The objective function is linear. All decision variables must be of power one and are only added or subtracted (no division or multiplication of decision variables is allowed).
- We have only one objective function. The goal is to find either a minimum or maximum solution.
- All constraints must be linear and only constraints of the form $\leq$, $\geq$, or $=$ are allowed.

3.2.2 Simplex Method

The Simplex method, which was developed by Dantzig in 1947 (Dantzig, 1949, 1951, 1962), is an effective method for solving LP problems. The Simplex method takes as input an LP in standard form and returns an optimal solution. Although its worst-case time complexity is exponential (Klee and Minty, 1972), it is very efficient in practice (Smale, 1983; Kelner and Spielman, 2006).

When searching for optimal solutions of LPs, we can make use of the fact that a linear inequality splits an n-dimensional search space into two halves which are called *half-spaces*. One half-space contains feasible solutions, the other one infeasible solutions. The intersection of all feasible half-spaces forms the feasible region and is called a *polyhedron* if the feasible region is unbounded or *polytope* if the feasible region is bounded. A set S in an n-dimensional vector space $\mathbb{R}^n$ is called a *convex set* if the line segment joining any pair of points of S lies entirely in S. When using linear constraints, the feasible region of any LP is always a convex set. The feasible, convex region forms a *simplex* which is the simplest possible polytope in a space of size n (the name simplex comes from the simplest possible polytope). A convex polytope can be defined either as the convex hull of a feasible region, or as the intersection of a finite set of half-spaces (Weyl, 1935). Solutions that are on the border of the feasible region (simplex) are called *boundary points*. Solutions that are feasible and no boundary points are called *interior points*. Feasible points that are boundary points and lie on the intersections of n half-spaces are called *corner points* or vertices of the simplex. The number of corner points grows exponentially with the size n of the problem.

The set of solutions for which the objective function obtains a specific value is a hyperplane of dimension $n - 1$. Because the set of feasible solutions is a convex set (simplex) and the objective function is linear, one of the corner points of the simplex is an optimal solution for an LP (if a bounded optimal solution exists). Consequently, optimal solutions can be found by examining all vertices of the simplex and choosing the one(s) where the objective function becomes optimal.

We give two examples. Example 1 is a two-dimensional problem, where the objective function $f(x) = x_1 + 2x_2$ must be maximized. We assume that $x_1, x_2 \geq 0$. In addition, there are three inequalities $x_2 - 2x_1 \leq 1$, $x_2 \leq 3$, and $3x_1 + x_2 \leq 9$. Figure 3.3(a) shows the different half-spaces and the resulting simplex (shaded area). One of the five vertices of the simplex is an optimal solution. The vertex $(2,3)$ is the optimal solution as the objective function becomes maximal for $f(2,3) = 8$. Figure 3.3(b) shows the same LP including the objective values. On the ground plane, we see the feasible region defined by the five different inequalities. The surface shows the objective function which becomes maximal for $(2,3)$.

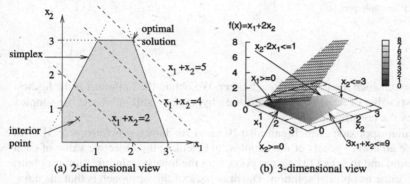

(a) 2-dimensional view   (b) 3-dimensional view

Fig. 3.3 LP example with two variables

Example 2 is a three dimensional $(x = (x_1, x_2, x_3)^T)$ minimization problem in canonical form (3.5):

$$\min \quad c^T x, \qquad (3.7)$$
$$\text{subject to} \quad Ax \geq b, \ x_i \geq 0,$$

where $c = \begin{pmatrix} 1 \\ 1 \\ -3 \end{pmatrix}$, $A = \begin{pmatrix} 1 & 0 & 1 \\ 0 & -1 & 1 \\ -1 & 0 & 1 \\ 0 & 1 & 1 \\ 0 & 0 & 1 \end{pmatrix}$, and $b = \begin{pmatrix} 2 \\ -1 \\ -1 \\ 2 \\ 1 \end{pmatrix}$

The set of five inequalities forms a polyhedron (unbounded simplex) with four corner points at $(1,1,1)$, $(1,2,1)$, $(2,2,1)$, and $(2,1,1)$. Figure 3.4 shows the five different constraints splitting the search space into a feasible and infeasible region. As the set of feasible solutions forms a convex set and the objective function is linear, the optimal solution is one of the four corner points. We plot the objective function $f(x) = x_1 + x_2 - 3x_3 = 1$ and all points in this plane have the same fitness value $f(x) = 1$. The optimal solution $(2,2,1)$ is contained in this plane and lies at the corner that is formed by the constraints $x_3 \geq 1$, $-x_1 + x_3 \geq -1$, and $x_2 + x_3 \geq 2$.

In general, we can find optimal solutions for LPs of size $n \leq 3$ using a graphical procedure. In a first step, we plot the different constraints in a two-dimensional or three-dimensional plot. If we assume that a non-empty, bounded feasible region (simplex) exists, then the optimal solution is one of the corner points of this simplex. To find the optimal solution, we have to consider the hyperplanes that are defined

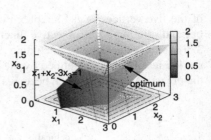

Fig. 3.4 LP example problem
with three variables

by $f(x) = c$, which are parallel to each other. We obtain the optimum at the highest
(or lowest) value of c such that the resulting hyperplane still intersects the simplex
formed by the constraints.

A similar approach (which can also be used for larger problems $n \geq 3$) is to
enumerate all corner points of the simplex, to calculate the objective value of each
corner point, and to select the corner point with the lowest (minimization problem)
objective value as optimal solution. The drawback of this approach is that the number of corner points that have to be evaluated increases exponentially with n.

A more systematic and usually more efficient approach is the Simplex method.
The Simplex method starts at some vertex of the simplex and performs a sequence of
iterations. In each iteration, it moves along an edge of the simplex to a neighboring
vertex with higher or equal objective value. It terminates at a local optimum which
is a vertex where all neighboring vertices have a smaller objective value. As the
feasible region is convex and the objective function is linear, there is only one local
optimum which is also the global optimum. The Simplex method only moves on the
convex hull of the simplex and makes use of the fact that the optimal solution of LPs
is never an interior point but always a corner point on the boundary of the feasible
region.

Before we can apply the Simplex method, we have to transform an LP from
canonical form (3.5) to a standard form (3.6) where all $b_i \geq 0$. Therefore, in a first
step, all constraints where $b_i \leq 0$ must be multiplied by -1. Then, we have to introduce slack and surplus variables for all inequality constraints. Slack variables are
introduced for all inequalities of the form $f(x) \leq b$ and surplus variables are used
for all inequalities $f(x) \geq b$. After these operations, the LP is in standard form and
one additional variable (either slack or surplus variable) has been introduced for
each inequality constraint. Then, a new variable x_i, called an *artificial variable* is
added to the left-hand side of each equality constraint that does not contain a slack
variable. Consequently, each equality constraint will then contain either one slack
variable or one artificial variable. A non-negative initial solution for this problem is
obtained by setting each slack and artificial variable equal to the right-hand side of
the equation in which it appears and setting all other variables, including the surplus
variables, equal to zero. All variables that are set unequal to zero form a basis and

are called basic variables. Usually, a basis contains m different variables, where m is the number of constraints ($m \geq n$).

How do slack, surplus, and artificial variables change the LP? Slack and surplus variables do not affect the constraints or the objective function. Therefore, they are incorporated into the objective function with zero coefficients. However, artificial variables change constraints because they are added to only one side of an equality equation. Therefore, the new constraints are equivalent to the original ones if and only if the artificial variables are set to zero. To guarantee such an assignment in the optimal solution (not in the initial solution), artificial variables are incorporated into the objective function with large positive coefficients M for minimization problems and large negative coefficients $-M$ for maximization problems.

We study this procedure for our example problem (3.7, p. 53). By introducing slack (x_5 and x_6) and surplus (x_4, x_7, and x_8) variables, we can remove the inequality constraints and transform the problem into standard form:

$$
\begin{array}{rcl}
x_1 \quad +x_3 -x_4 & = & 2 \\
x_2 -x_3 \quad +x_5 & = & 1 \\
x_1 \quad -x_3 \quad +x_6 & = & 1 \\
x_2 +x_3 \quad -x_7 & = & 2 \\
x_3 \quad -x_8 & = & 1
\end{array}
\tag{3.8}
$$

Then, we have to add additional artificial variables to all constraints that do not contain a slack variable. Therefore, we add the artificial variables x_9 and x_{10} to the first and last constraint, respectively. After introducing the artificial variables we get the equations

$$
\begin{array}{rcl}
x_1 \quad +x_3 -x_4 \quad +x_9 & = & 2 \\
x_2 -x_3 \quad +x_5 & = & 1 \\
x_1 \quad -x_3 \quad +x_6 & = & 1 \\
x_2 +x_3 \quad -x_7 \quad +x_{10} & = & 2 \\
x_3 \quad -x_8 \quad +x_{11} & = & 1
\end{array}
\tag{3.9}
$$

We get an initial, but infeasible, solution $x = (0,0,0,0,1,1,0,0,2,2,1)^T$ by setting all decision variables (x_1, x_2, and x_3) as well as all surplus variables (x_4, x_7, and x_8) to zero. The solution is infeasible as some constraints on the decision variables are violated. We can calculate the objective value of this infeasible solution using $c^T = (1,1,-3,0,0,0,0,0,M,M,M)$. We see that the artificial variables are considered in the objective function with a large positive coefficient M. This ensures that the artificial variables are set to zero for the optimal solution. At the end of the first step, the variables x_5, x_6, x_9, x_{10}, and x_{11} form the initial basis.

After obtaining an initial solution (which is usually infeasible if not all decision variables are in the basis), the Simplex method iteratively moves to an adjacent extreme point by exchanging one of the variables that is in the basis for a variable that is not in the basis. Changing the basis means that one of the variables currently in the basis is set to zero and, in exchange, one of the current non-basis variables is selected to be in the basis. The main differences between different variants of the Simplex method are which variable is brought into the basis and which variable is removed from the basis. Usually, a non-basic variable is brought into the basis which makes the solution least infeasible or leads to the highest improvement in the objective function. The basic variable to leave the basis is the one which is most infeasible or expected to go infeasible first. Usually, the rules for selecting the variables that enter and leave the basis are heuristics.

By performing iterative basis changes, the Simplex method moves from one solution to the next, selecting the new basic variables such that the objective function becomes lower (minimizing). If a feasible solution exists, the Simplex method finds it. We know that we have found the optimal solution if no basis change leads to a further reduction of the objective function. The Simplex method is fast as the optimal solution is often reached after only $2m \ldots 3m$ basis changes.

If the initial solution is infeasible, often a two-phase approach is used. In the first phase, the entering and leaving variables are chosen such that a feasible solution is obtained. The second phase starts after a feasible solution has been found. Then, the Simplex method tries to improve the solution. We should have in mind that to obtain a feasible solution usually all decision variables must be in the basis.

For more details on the functionality of the Simplex method and appropriate strategies of moving from one corner of the simplex to adjacent corners, the interested reader is referred to the literature. Basically each textbook on OR or linear programming describes the Simplex method in detail and provides additional background information. Bronson and Naadimuthu (1997) or Grünert and Irnich (2005, in German) provide an illustrative introduction but also other literature is recommended (Papadimitriou and Steiglitz, 1982; Chvatal, 1983; Williams, 1999; Cormen et al, 2001; Hillier and Lieberman, 2002; Winston, 1991; Domschke and Drexl, 2005).

3.2.3 Simplex and Interior Point Methods

Prior to 1979, the Simplex method was the dominating and main solution approach for LPs as it solved routinely and efficiently even very large LPs. The Simplex method showed the nice property that, on almost all relevant real-world problems, the number of iterations that are necessary to move to the feasible optimal solution is a small multiple of the problem dimension n. However, as problem instances exist where the Simplex method visits every vertex of the feasible region (Klee and Minty, 1972), the worst-case complexity of the Simplex method is exponential in the problem dimension n.

The situation changed in 1979, when Lenid Khachian proposed an ellipsoid method (Khachian, 1979) which was the first polynomial-time LP algorithm. This method is a specific variant of general non-linear approaches (Shor, 1970; Yudin and Nemirovskii, 1976) as it does not search through the convex hull of the feasible region but inscribes a sequence of ellipsoids with decreasing volume in the feasible region. Such methods that do not search on the convex hull but approach the optimal solution (which is a corner-point) from within the feasible region are called *interior point methods*. Although the algorithm had a great impact on theory, its impact on practice was low as the algorithm usually always reaches the worst-case bound and needs on average a larger number of search steps than the Simplex method. Khachian's method is a nice example of a polynomial optimization method that on average needs more search steps than a method with worst-case exponential behavior. However, it was the first approach to succeed in establishing a connection between linear problems and non-linear problems where the objective function depends non-linearly on the decision variables. Before, these two types of optimization problems were solved with completely different methods although one is a strict subset of the other.

In 1984, Karmarkar presented a more advanced and faster polynomial-time method for solving LPs (Karmarkar, 1984). In this method, a sequence of spheres is inscribed in the feasible region such that the centers of the spheres converge to the optimal solution. There is a formal equivalence between this method and the logarithmic barrier method (Gill et al, 1986; Forsgren et al, 2002). Comparing the interior point method from Karmarkar to the Simplex method revealed similar or better performance of interior point methods for many problem instances (Gill et al, 1986). For more information on the development of interior point methods and a detailed explanation of their functionality, we refer the interested reader to Wright (1997) and Wright (2005).

In the last few years, a number of improvements have been developed for the Simplex method as well as interior point methods. Nowadays, the performance differences between these two approaches depend strongly on the particular geometric and algebraic properties of the LP problem. In general, however, interior-point methods show similar or better performance than Simplex methods for larger problems when no prior information about the optimal solution is available. In contrast, when prior information is available and a so called "warm start" is possible (starting from a previously found solution), the situation is reversed and Simplex methods tend to be able to make much better use of it than interior-point methods (Maros and Mitra, 1996).

Prior to 1987, all commercial LP problem solvers used the Simplex algorithm. Nowadays, there are also a number of free interior-point implementations available that are competitive with commercial variants. For an overview of current interior-point methods we refer to Benson (2010a) and Benson (2010b). As the emergence of interior point methods spurred the development of more efficient Simplex methods, there are a number of free and commercial LP solvers based on the Simplex method available (Mittelmann, 2010).

Summarizing, there are efficient methods available for linear, continuous, optimization problems. Linear optimization problems belong to the class P as they can be solved in polynomial time using interior point methods. The Simplex method, which has worst-case exponential running time, and interior point methods, which have worst-case polynomial running time, show similar performance and can solve even large problem instances in a short time. The main limitation of linear optimization methods is that they can only be applied to linear optimization problems where the objective function and the constraints are linear. If the objective function or one of the constraints is not linear, LP optimization methods cannot be applied. When creating an LP model for a real-world optimization problem, we must make sure that the character of the problem is described well by a linear model. Otherwise, we may find an optimal solution for the LP model but the solution does not fit the underlying non-linear optimization problem.

3.3 Optimization Methods for Linear, Discrete Problems

In the previous section, we studied optimization methods for linear problems where the decision variables are continuous. This section describes optimization methods for problems where all decision variables are integers. Such problems are also called *combinatorial problems* if the number of possible solutions is finite and we are able to enumerate all possible solutions. Discrete decision variables are used to model a variety of different combinatorial problems like assignment problems (e.g. time tabling), scheduling problems (e.g. traveling salesman problem, job scheduling problems, or Chinese postman problem), grouping problems (e.g. cutting and packing problems or lot sizing problems), and selection problems (e.g. knapsack problems, set partitioning problems, or set covering problems).

Representative optimization methods for combinatorial optimization problems are decision tree-based enumeration methods and cutting plane methods. The use of decision trees allows us to formulate the process of problem solution as a sequence of decisions on the decision variables of the problem. Common methods working with decision trees are uninformed and informed graph search algorithms like depth or A*-search, branch-and-bound approaches, or dynamic optimization. Cutting plane methods are often based on the Simplex method and add additional constraints (cuts) to a problem to ensure that the decision variables in the optimal solution are discrete. The optimization methods presented in this section can not only be used for linear problems but can also be applied to non-linear optimization problems.

Section 3.3.1 introduces integer linear problems. This is followed by representative decision tree methods. In particular, we describe uninformed and informed tree search methods (Sect. 3.3.2), branch-and-bound methods (Sect. 3.3.3), and dynamic programming (Sect. 3.3.4). An overview of the functionality of cutting plane methods is given in Sect. 3.3.5.

3.3.1 Integer Linear Problems

Integer linear programs (ILP) are LPs where the decision variables are integers. ILPs in canonical form can be formulated as

$$\max \quad c^T x \tag{3.10}$$
$$\text{subject to} \quad Ax \leq b,$$
$$x_i \in \mathbb{N}_0,$$

where $\mathbb{N}_0$ is the set of non-negative integers $x \geq 0$. Problems are called *mixed integer linear programs* (MILP) if there are discrete decision variables x_i and continuous decision variables y_j. Their canonical form is

$$\max \quad c^T x + d^T y \tag{3.11}$$
$$\text{subject to} \quad Ax + By \leq b,$$
$$x_i \in \mathbb{N}_0,$$
$$y_j \geq 0.$$

Both types of problems are discrete optimization problems where some (MILP) or all (ILP) of the decision variables are not allowed to be fractional.

In general, models using integer variables can be converted to models using only binary variables $x_i \in \{0, 1\}$. Sometimes, such a conversion provides some nice properties which can be exploited, but typically more variables are needed to characterize the same integer problem.

If we drop the integer constraints of an ILP, we get an LP in canonical form which is called the *relaxation* of the ILP or the relaxed problem. We can apply standard methods for continuous, linear problems like the Simplex method to solve relaxed ILPs. However, usually the optimal solution of a relaxed problem is not integral and is, thus, not a feasible solution of the underlying ILP.

However, applying LP optimization methods to the relaxation of an ILP and ignoring the integer feasibility constraint can give us a first bound on the solution quality of the underlying ILP. Even more, if the optimal solution obtained for the LP is integral, then this solution is also the optimal solution to the original ILP. However, usually some or all variables of the optimal solution for the relaxed problem are fractional and the optimal solution lies in the interior of the resulting simplex. Straightforward approaches to obtain an integral feasible solution from a fractional solution are

- rounding or
- search in the neighborhood of the optimal solution.

Rounding may be a useful strategy when the decision variables are expected to be large integers and insensitive to rounding. However, rounding can also be completely misleading as many integer problems are not just continuous problems with additional integrality constraints, but integer constraints are used to model combinatorial constraints, logical constraints, or non-linearities of any sort. Due to the nature of such ILPs, rounding often does not work as it would defeat the purpose of the ILP formulation (Papadimitriou and Steiglitz, 1982, Chap. 13). The same holds true for local search in the neighborhood of the optimal solution. Often the search space is highly constrained and searching around the optimal solution of the relaxed problem does not yield the optimal solution of the ILP.

Although solving the relaxed problem usually does not allow us to find the integral, optimal solution, it gives us a lower (for minimization problems) bound on the objective value of the optimal solution of an ILP. The gap between the objective values of the optimal ILP solution and the optimal relaxed solution is called the *integrality gap*. Obtaining a lower bound on the objective value of the optimal solution can be helpful to estimate how close we are to the optimal solution. For example, if we have found a feasible solution for an ILP and know that the gap between the quality of this solution and the relaxed solution is small, spending much effort in finding better solutions may not be necessary as the potential improvement is low.

We want to give two examples where rounding or local search around the optimal solution of the relaxed problem does not find the optimal solution. In the first example, we maximize the objective function $f(x) = 6x_1 + x_2$. There are additional constraints $x_1 \geq 0$, $x_2 \geq 0$, $x_2 \leq 4$, and $5x_1 + x_2 \leq 19$. We are searching for an integral solution, where $x_i \in \mathbb{N}_0$. Figure 3.5(a) shows the problem. All feasible solutions of the ILP are denoted as dots. The optimal solution of the relaxed problem is $(3.8, 0)$ with $f(3.8, 0) = 23$. Rounding this solution would yield $(4, 0)$, which is an infeasible solution. The nearest integral feasible solution (using Euclidean distances) is $(3, 0)$ with $f(3, 0) = 18$. This is not the optimal solution as the correct, optimal, solution of the ILP is $(3, 4)$ with $f(3, 4) = 22$.

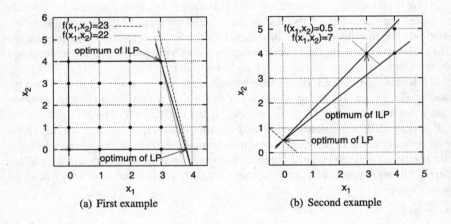

(a) First example   (b) Second example

Fig. 3.5 Two two-dimensional ILP example problems

The second example is a problem where finding the optimal solution of the relaxed problem is not at all helpful for solving the original ILP (Papadimitriou and Steiglitz, 1982; Bonze and Grossmann, 1993). We have an ILP of the form

$$\min \quad x_1 + x_2 \qquad (3.12)$$
$$\text{subject to} \quad Ax \geq b,$$
$$x_1, x_2 \in \mathbb{N}_0,$$

$$\text{where } A = \begin{pmatrix} (1-2n) & 2n \\ (2n-1) & 2(1-n) \end{pmatrix} \text{ and } b = \begin{pmatrix} n \\ 1-n \end{pmatrix}.$$

Figure 3.5(b) shows the resulting feasible solutions for $n = 4$. Independently of n, the optimal solution of the relaxed problem is $(0, 0.5)$. The optimal solution of the ILP is $x^* = (n-1, n)^T$ with $f(x^*) = 2n - 1$. With increasing n the integrality gap increases. Both examples illustrate that rounding the optimal solution of an LP or searching in the neighborhood of this solution does not necessarily yield the optimal solution for the original ILP.

3.3.2 Uninformed and Informed Search

We can use graphs and trees to model combinatorial search spaces if the size of the search spaces is limited. Each node of the graph represents a possible solution of a combinatorial optimization problem and an edge between two nodes indicates some relationship. Common relationships between nodes are the similarity of solutions with respect to a distance metric (see Sect. 2.3.1). For example, two solutions x and y are related and an edge exists between them if their Hamming distance $d(x, y) = 1$.

Formally, a graph $G = (V, E)$ consists of a set V of $n = |V|$ vertices (solutions) and a set E of $m = |E|$ vertex pairs or edges (relationships). A path is a sequence of edges connecting two nodes. A graph is connected if there is a path between every two nodes. Trees are a special variant of graphs where there is exactly one path between every two nodes. Therefore, a tree T is defined as a connected graph with no cycles. For a tree T with n nodes, there are exactly $n - 1$ links.

Figure 3.6(a) uses a graph to describe the relationships between the different solutions of a combinatorial optimization problem with three variables $x_i \in \{0, 1\}$. There are $2^3 = 8$ different solutions and we can use a graph to model the search space (analogously to Sect. 2.3.3, p. 21). As each node represents a possible solution, the number of solutions equals the number of nodes.

Instead of a graph, we can use a hierarchical tree of depth d to represent a search space. One node is selected to be the root node. Figure 3.6(b) models the example using a hierarchical tree. An undirected edge between x and y indicates neighboring solutions (Hamming distance $d(x, y) = 1$). The depth d is equivalent to the maximum distance $\max_{x \in X}(d(x, r))$ between a solution x and the root node. When using a hierarchical tree, the same solutions can occur multiple times in the tree as no cycles are allowed in a tree.

We can also use tree structures to model the process of iteratively assigning values to the n decision variables. Then, the maximum depth d of a tree is equal to n. At each level $i \in \{1, \ldots, n\}$, we assign possible values to the ith decision variable.

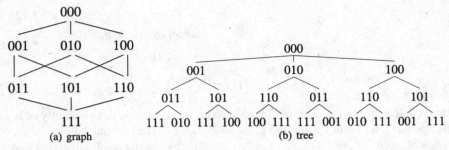

Fig. 3.6 A combinatorial search space with $x \in \{0,1\}^3$

Tree structures modeling such a decision process are called *decision trees*. Figure 3.7 shows a possible decision tree for the example from above. In the root node, all three variables are undecided and we successively assign values to the three binary decision variables starting with x_0. On the lowest level we can find all possible solutions for the problem.

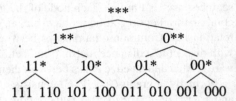

Fig. 3.7 Decision tree

The examples demonstrate that we can use tree structures to describe solution strategies for discrete optimization problems if the number of options for the decision variables is limited. We can use trees

- either as hierarchical trees where each node represents a complete solution of the problem (see Fig. 3.6(b)) or
- as decision trees where we assign on each level possible values to a decision variable. Solutions are completed at depth n of the tree (see Fig. 3.7).

When representing solutions of a combinatorial optimization problem using a tree structure, we can apply graph search (or graph traversal) algorithms to systematically go through all the nodes in the tree, often with the goal of finding a particular node, or one with a given property. We can distinguish between

- *uninformed search methods* which completely enumerate all possible solutions using a fixed search behavior and
- *informed search methods* that are problem-specific and use the estimated distance to the goal node to control the search.

In contrast to unordered linear structures where we can start at the beginning and work through to the end, searching a tree is more complex as it is a hierarchical structure modeling relationships between solutions.

3.3.2.1 Uninformed Tree Search Strategies

Graph search algorithms are uninformed if no problem-specific information guides the search. The search through a tree follows a predefined pattern which enumerates all possible solutions. Therefore, search strategies differ by the order in which they expand the nodes of a tree, where a node expansion is equivalent to examining all adjacent nodes of a lower level. Common uninformed graph search algorithms are

- breadth-first search,
- depth-first search, and
- uniform-cost search.

Breadth-first search starts with the root node and successively expands the nodes in order of creation. Figure 3.8(a) illustrates the order in which breadth-first search expands the nodes. The search terminates if a node is the goal (e.g. optimal solution). With the depth d of a hierarchical tree and defining b as the average branching factor (average number of expanded nodes), time and space complexity is $O(b^d)$.

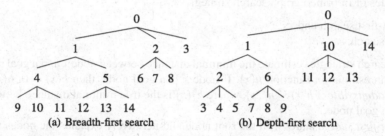

(a) Breadth-first search   (b) Depth-first search

Fig. 3.8 Order of node expansion for breadth-first search and depth-first search

Depth-first search also starts with the root node and expands the last-created node. Therefore, depth-first search moves directly to the bottom of the tree and then successively expands the remaining nodes. Figure 3.8(b) shows the order in which the nodes are expanded. The time complexity of depth-first search is $O(b^d)$. Space complexity is lower ($O(bd)$) than breadth-first search as breadth-first search traverses the tree from "left to right" and not from "top to bottom".

Uniform-cost search is an uninformed search method which uses a cost that is associated to each node for controlling the search. Uniform-cost search iteratively expands the node with the lowest associated cost. It behaves like breadth-first search if the cost of a node is equal to its distance to the root node. Often, costs are assigned to edges and the cost of a node is the sum of the costs of the edges on the path between the node and the root node. Figure 3.9 shows the order in which the nodes

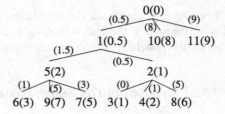

Fig. 3.9 Order of node expansion for uniform-cost search. Edge and node costs are in brackets

are expanded (the numbers in brackets show the cost of each node and edge, respectively). The space and time complexity of uniform-cost search is $O(b^d)$.

Other examples of uninformed graph search methods are modifications of depth-first search like iterative-deepening search or depth-limited search and bidirectional search (for details see for example Russell and Norvig (2002) or Zhang (1999)).

3.3.2.2 Informed Tree Search Strategies

Informed search strategies use an *evaluation function h* to guide the search through a tree. $h(x)$ measures or estimates the cost of the shortest path between node x and a goal node. Finding a path from the root to a goal node is easy if h is accurate as then we only need to iteratively expand the nodes with minimal h. Representative examples of informed graph search strategies are

- best-first search and
- A* search.

Both search strategies estimate the minimal distance between node x and a goal node using an estimation function $h(x)$. If node x is a goal node, then $h(x) = 0$. $h(x)$ is called *admissible* if $h(x) \leq h^*(x)$, where $h^*(x)$ is the true minimal distance between x and a goal node.

Best-first search starts with the root node and iteratively expands the nodes with lowest h. If h is accurate ($h(x) = h^*(x)$), the goal is found after d expansions (if we assume that a goal node is located at depth d). Best-first search is called *greedy search* (GS) if h estimates the minimum distance to a goal node.

A search* combines uniform-cost search with best-first search. The cost $f(x)$ of a node x is calculated as $f(x) = g(x) + h(x)$, where $g(x)$ is the distance between root node and x and $h(x)$ estimates the minimal distance between x and a goal node. A* always expands the node x with minimal $f(x)$ and stops after a goal node is found. The complexity of A* depends on the accuracy of the estimation h. If h is accurate, time complexity is $O(bd)$. However, in general h is inaccurate and time and space complexity can increase up to $O(b^d)$. If h is admissible, the first goal node that is found has minimal distance g to the root node.

In principle, there are two different possibilities for optimization methods to use decision trees (compare also Sect. 3.3.2): First, each node of a tree represents a problem solution (for an example, see Fig. 3.6(b), p. 62). This is, for example, the case for most modern heuristics which search through the search space by iteratively sampling solutions. Such optimization methods are interested in finding with minimal effort a node which represents an optimal solution. Expanding a node x usually means investigating all neighboring solutions of x. An evaluation function $h(x)$ can, for example, estimate the minimal distance d_{x,x^*} between node x and a goal node x^*, which represents an optimal solution. However, such an estimation is often difficult as we do not know where the optimal solution can be found. More useful evaluation functions would be, for example, based on the objective values or other properties of solutions.

Second, starting with a root node where no values are assigned to the decision variables, optimization methods can search by iteratively assigning values to the decision variables (for an example, see Fig. 3.7, p. 62). To obtain a solution, a method must traverse the tree down to depth d. In this case, edge costs describe the effect of setting a decision variable to a particular value. $h(x)$ measures the minimal cost of a path from a node x, where some of the decision variables are fixed, to an optimal solution, which can be found at depth d in the decision tree.

3.3.2.3 Example

We want to compare the behavior of the different search methods for the *traveling salesman problem* (TSP). The TSP is a well-studied and common NP-hard combinatorial optimization problem. It can be formulated as follows: given a collection of cities (nodes) and distances $d(i, j)$ between each pair of nodes i and j, the TSP is to find the cheapest way of visiting all of the cities exactly once and returning to the starting point. A TSP is called *symmetric* if $d(i, j) = d(j, i)$. The size of the search space grows super-exponentially, as for symmetric problems there are $(n-1)!/2$ different solutions. A solution can be represented as a permutation of the n cities, which is also called a *tour*.

We want to study a symmetric problem with four cities ($n = 4$) denoted a, b, c, and d. The distances $d(i, j)$ are shown in Fig. 3.10. There are only $3!/2 = 3$ different solutions (*adbca* with cost 12, *abdca* with cost 17, and *adcba* with cost 11).

Fig. 3.10 Distances between cities a, b, c, and d

We can model the search space using a decision tree where we start with an empty tour and successively assign a city to a tour. We assume that the tour always starts with city a. The cost of an edge between nodes i and j is equivalent to the distance $d_{i,j}$. Figure 3.11 shows the resulting tree. The numbers next to the edges are the edge costs. The node labels represent the selected city and the node number. The number in brackets is $g(x)$ (cost of path from root to node x). The optimal solution x^* is represented twice (*abcda* (node 16) and *adcba* (node 21)) and has cost $f(x^*) = d(a,b) + d(b,c) + d(c,d) + d(d,a) = 11$.

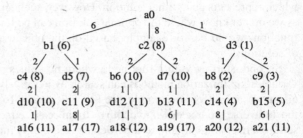

Fig. 3.11 Decision tree for the example TSP. The number in brackets is $g(x)$

We want to illustrate how different search strategies traverse the tree. Breadth-first search expands the nodes in the order of creation and expands the nodes in the order 0-1-2-3-4-5-6-7-8-9-10-11-12-13-14-15. Depth-first search expands the nodes in the order 0-1-4-10-5-11-2-6-12-7-13-3-8-14-9-15. Uniform-cost search considers the distance $g(x)$ between node x and root node. $g(x)$ measures the length of the partial tour represented by node x. Uniform-cost search expands the nodes in the order 0-3-8-9-14-15-1-5-4-2-11-6-7-10. The nodes 4 and 2 as well as 6, 7, and 10 have equal cost g and are, thus, expanded in random order. The nodes 12 and 13 need not be expanded since their cost g is equal to the cost of an already found solution.

Informed search methods estimate the minimal distance to a goal state. The estimation function h is accurate if it correctly predicts the length of the optimal tour. However, as we do not know the optimal solution, we need an estimate $h(x)$. We assume that $h(x)$ is the length of the shortest path from city x to city a containing no cities that are already part of the tour. In general, $h(x) \leq h^*(x)$. h provides an inaccurate estimation but it allows us to demonstrate the functionality of informed search methods.

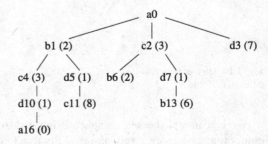

Fig. 3.12 Decision tree for solving the example TSP using greedy search. The number in brackets indicates $h(x)$ which is the length of the shortest tour from node x to a

Figure 3.12 illustrates the functionality of greedy search. It uses the node numbers introduced in Fig. 3.11. The numbers in brackets indicate $h(x)$. Greedy search starts with expanding node 0. We obtain $h(1) = 1 + 1 = 2$ (we go back to city a using the path bda), $h(2) = 2 + 1 = 3$ (we go back to a using cda), and $h(3) = 1 + 6 = 7$ (we go to a using path dba). Consequently, we continue with expanding node 1 and get $h(4) = 3$ (we use path cda) and $h(5) = 1$ (we directly go back to a from d). We continue with expanding node 5 and get $h(11) = 8$. Next, we expand nodes 4 and 2 to obtain node 10 with $h(10) = 1$ and the nodes 6 and 7 with $h(6) = 2$ and $h(7) = 1$. Finally, we expand nodes 10 and 7 to obtain node 16 with $h(16) = 0$ and node 13 with $h(13) = 6$. Therefore, we have found a solution (node 16), terminate the search, and return $abcda$. Although in this example, greedy search returns the optimal solution, this is not always the case. Greedy search is myopic and often gets stuck in local optima, not returning an optimal solution.

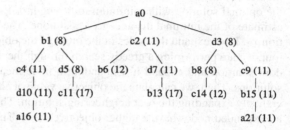

Fig. 3.13 Decision tree for solving the example TSP using A* search. The numbers in brackets indicate $f(x)$

Figure 3.13 illustrates the functionality of A* search. The numbers in brackets indicate $f(x) = g(x) + h(x)$. A* search starts with expanding node 0 and we obtain $f(1) = 6 + 2 = 8$, $f(2) = 8 + 3 = 11$, and $f(3) = 1 + 7 = 8$. We continue expanding nodes 1 and 3 and obtain nodes 4, 5; 8, and 9 with $f(4) = 8 + 3 = 11$, $f(5) = 7 + 1 = 8$, $f(8) = 2 + 6 = 8$ and $f(9) = 3 + 8 = 11$. Expanding nodes 5 and 8 yields nodes 11 and 14 with $f(11) = 9 + 8 = 17$ and $f(14) = 4 + 8 = 12$. We continue with expanding the nodes with lowest f and expand nodes 2, 4, and 9. All have $f = 11$. We obtain nodes 6, 7, 10, and 15 with $f(6) = 10 + 2 = 12$, $f(7) = 10 + 1 = 11$, $f(10) = 10 + 1 = 11$, and $f(15) = 5 + 6 = 11$. Continuing with nodes 7, 10, and 15, which all have $f = 11$, we obtain the nodes 16, 13, and 21 with $f(16) = 11 + 0 = 11$, $f(13) = 11 + 6 = 17$, and $f(21) = 11 + 0 = 11$. Since nodes 16 and 21 are goal nodes (the tour is completed) and there are no other nodes with lower f, we can terminate the search and return $abcda$ and $adcba$ as optimal solutions. We must have in mind that A* guarantees finding the optimal solution.

We can use tree structures not only for describing the process of assigning values to the decision variables (where nodes can also represent incomplete solutions) but also for representing a search space containing only complete solutions. Then, nodes represent complete tours and an edge indicates similarity between two tours. Figure 3.14 shows the resulting search tree using tour $abcda$ as root node. We assume that each tour starts and ends at city a. The edges indicate neighboring solutions. x and y are neighbors if their Hamming distance $d(x, y) = 2$; this means exchanging the position of two cities in a tour results in a neighboring solution. Given n cities, the maximum depth d of the tree is $n - 2$. The branching factor b of a node depends on the level of the node and decreases with increasing tree depth.

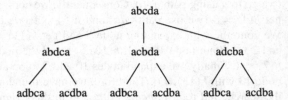

Fig. 3.14 Search tree for the example TSP containing only complete solutions

We can use standard uninformed and informed tree search methods to traverse the tree. Uninformed search methods just enumerate all possible solutions and return the optimal solution with minimum tour length. Informed search methods need an estimate of the minimal distance to a goal node. The goal node is the optimal solution $x*$. If we expand the nodes in the order of the objective value of the represented tour, we are performing a greedy search through the search space. We start at a random solution and expand it. Expanding a node means generating all neighboring solutions. Then, we calculate the objective value of all neighbors and continue by iteratively expanding the best neighboring solution. The search stops at a node where all expanded nodes have a higher objective value. This node is a local optimum. In our example, greedy search starts with expanding node *abcda*. Since all expanded nodes have the same (*adcba*) or higher objective value, greedy search returns either *abcda* or *adcba* as optimal solutions.

3.3.3 Branch and Bound Methods

Besides complete enumeration of the search space, which can be performed using the search strategies discussed in the previous section, a common enumerative approach is *branch-and-bound*. The general idea behind branch-and-bound algorithms, which can also be used for non-linear problems, is to recursively decompose a problem into subproblems. For example, when solving ILPs, we can obtain subproblems by fixing or introducing additional constraints on the decision variables. By subsequently adding constraints to the original problem, we get subproblems and relaxed versions of the subproblems are solved using LP methods. The process of subsequently adding additional constraints to the original problem is called *branching* and can be modeled using hierarchical tree structures.

Bounding refers to removing (also called killing or fathoming) subproblems from further consideration. Subproblems that have been killed are not considered any more and are not decomposed into subproblems. Subproblems are killed if a bound (e.g., obtained by solving the relaxed version of the subproblem) is below an existing lower bound (maximization problem). For maximization problems, the objective values of feasible integral solutions can be used as lower bounds on the objective value.

Branch-and-bound methods were introduced by Land and Doig (1960) and the first practical implementation was presented by Dakin (1965). Detailed descriptions of branch-and-bound methods can be found in standard OR and optimization literature (Papadimitriou and Steiglitz, 1982; Bronson and Naadimuthu, 1997; Hillier and Lieberman, 2002). Commercial branch-and-bound implementations for solving ILPs start with solving the relaxed version of the original ILP using LP solving methods like the Simplex method (see Sect. 3.2.2). The solution of the relaxed problem is the optimal solution of the ILP if it is integral. The original ILP is infeasible if the relaxed problem cannot be solved. Otherwise, at least one of the integer variables of the relaxed optimal solution is fractional. Commercial solvers usually use one fractional variable and create two subproblems such that the fractional solution is excluded but all feasible integer solutions still remain feasible. These new ILPs with an additional constraint represent nodes in a branching tree. The nodes in the search tree are iteratively expanded by adding additional constraints to the subproblems and solving the relaxed versions of the subproblems. Nodes can be killed if the solution to a subproblem is infeasible, satisfies all integrality restrictions, or has an objective function value worse than the best known integral solution.

Therefore, we can formulate the branch-and-bound method for solving ILPs (maximization problems) as follows:

1. Solve the linear relaxation of the original ILP. This gives an upper bound on the objective value of the optimal solution. The lower bound is set to $-\infty$. If the obtained solution is integral, stop, and return the optimal solution.
2. Create two new subproblems by branching on a fractional variable. Solving the linear relaxations of the two subproblems returns upper bounds on the two subproblems. A subproblem can be *killed* (or fathomed) when any of the following occurs:

   - All variables in the solution for the relaxed subproblem are integral. If the objective value is larger than the existing lower bound, it replaces the existing lower bound.
   - The relaxed subproblem is infeasible.
   - The objective value of the fractional solution is below the current lower bound. Then, the subproblem is killed by a bounding argument.

3. If any subproblems exist that are not yet killed (we call such subproblems active), choose one and continue with step 2. Otherwise, stop and return the current lower bound as the optimal solution.

Branch-and-bound is an exhaustive search method and yields the optimal solution to an ILP if it exists. The maximum depth of the resulting decision tree is n, where n is the number of decision variables. Therefore, the time complexity of branch-and-bound is exponential ($O(2^n)$) as, in the extreme case, we have to branch on all n decision variables. When implementing branch-and-bound methods, there are three important aspects:

- branching nodes,
- bounding solutions, and
- traversing the tree.

We briefly discuss these three aspects in the following paragraphs. *Branching* partitions a set of solutions into two mutually exclusive sets. When using search trees, each subset in the partition is represented by a child of the original node. For ILPs, linear relaxation of the problem can result in a solution with a non-integral variable x_i^*. Branching on the variable x_i^* creates two mutually exclusive ILPs with either the constraint $x_i \leq \lfloor x_i^* \rfloor$ or the constraint $x_i \geq \lceil x_i^* \rceil$. This branching step shrinks the feasible region such that the current non-integral solution for x_i is eliminated from further consideration but still all possible integral solutions to the original problem are preserved.

For *bounding*, we need an algorithm that calculates a lower bound on the cost of any solution in a given set. In general, a bound can be obtained by calculating any feasible solution of the original problem. For ILPs, we obtain a lower bound by finding a feasible, integral solution. All ILPs whose relaxed problems yield values of the objective function lower than the lower bound can be discarded. All such problems are excluded from further consideration.

We can order the different ILPs using a *search tree* with the original ILP as a root node. Branching creates two child nodes and bounding discards branches of the tree from further consideration. As usually more than one node of the search tree is active (this means it is not yet fathomed or killed), we need a search strategy for deciding which active node should be expanded next. For this task, we can use for example uninformed search methods described in Sect. 3.3.2. Other search strategies are to expand the node with the highest number of integral variables in the optimal solution or the node with the highest objective value. After choosing an active node to be expanded, we also have to choose a variable for which an additional constraint should be added (usually not only one variable is fractional). There are different strategies available (for an overview see Lee and Mitchell (2001) or Lee (2002)) but often the fractional variable is chosen that is furthest from being an integer (the fractional part is closest to 0.5).

The remainder of this section gives an example of how to use branch-and-bound to solve an ILP. We want to find the optimal solution for the following two-dimensional ILP:

$$\begin{aligned}
\max \quad & f(x) = 2x_1 + x_2 & (3.13)\\
\text{subject to} \quad & x_1 + 4x_2 \leq 9,\\
& 8x_1 + 2x_2 \leq 19\\
\text{with} \quad & x_1, x_2 \in \mathbb{N}_0 \text{ and } x_1, x_2 \geq 0.
\end{aligned}$$

Figure 3.15 plots the original ILP and indicates the optimal solution of the ILP and of the relaxed ILP. Solving the relaxed ILP (we drop the integrality constraints and use the Simplex method for the resulting LP), we find the optimal fractional solution

$x^* = (\frac{29}{15}, \frac{53}{30})$ with $f(x^*) = 5\frac{19}{30}$. Therefore, we have an upper bound for the optimal solution.

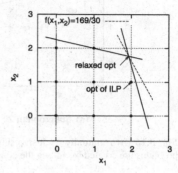

Fig. 3.15 Original ILP

In a first step, we create two mutually exclusive subproblems of (3.13) by introducing additional constraints on one of the two variables. In the optimal solution for the relaxed problem, both variables x_1 and x_2 are fractional. Because x_1 is closer to a feasible integral solution $(\min((x_2 - \lfloor x_2 \rfloor), (\lceil x_2 \rceil - x_2)) > \min((x_1 - \lfloor x_1 \rfloor), (\lceil x_1 \rceil - x_1)))$, we choose x_2 and define additional constraints on x_2. As $1 < x_2^* < 2$, we can create two mutually exclusive subproblems by adding either the constraint $x_2 \leq 1$ or the constraint $x_2 \geq 2$. By splitting up the search space in such a way, no feasible integral solutions are excluded. Therefore, we get the two mutually exclusive subproblems:

Subproblem 2

$$\max \quad 2x_1 + x_2 \qquad (3.14)$$
$$\text{subject to} \quad x_1 + 4x_2 \leq 9,$$
$$8x_1 + 2x_2 \leq 19$$
$$x_2 \geq 2$$
$$\text{with} \quad x_1, x_2 \in \mathbb{N}_0 \text{ and } x_1, x_2 \geq 0.$$

Subproblem 3

$$\max \quad 2x_1 + x_2 \qquad (3.15)$$
$$\text{subject to} \quad x_1 + 4x_2 \leq 9,$$
$$8x_1 + 2x_2 \leq 19$$
$$x_2 \leq 1$$
$$\text{with} \quad x_1, x_2 \in \mathbb{N}_0 \text{ and } x_1, x_2 \geq 0.$$

Figure 3.16 shows the two resulting subproblems. The optimal solution x^* of the relaxed subproblem 2 is integral with $x_1 = 1$ and $x_2 = 2$. The objective value of the optimal solution is $f(x^*) = 4$. As the optimal solution for the relaxed subproblem 2 is integral, we have a lower bound on the solution quality. Therefore, all other active subproblems where the optimal solution of the relaxed problem has a lower objective value can be killed and removed from further consideration. In our case, no other subproblems can be killed as the optimal solution of the relaxed subproblem 3 is $x^* = (2\frac{1}{8}, 1)$ with $f(x^*) = 5.25 > 4$.

As subproblem 3 is the only active problem (subproblem 2 is not active any more as the optimal solution for the relaxed problem is integral), we have to split it up into

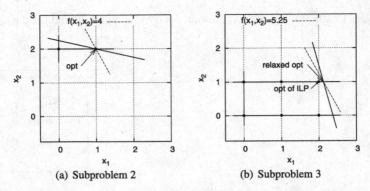

(a) Subproblem 2   (b) Subproblem 3

Fig. 3.16 Two mutually exclusive subproblems created from the original problem (3.13). The problems are created by adding the constraints $x_2 \geq 2$ and $x_2 \leq 1$

two exclusive subproblems. The only fractional variable of the optimal solution of the relaxed subproblem 3 is $x_1^* = 2\frac{1}{8}$. Therefore, we create two new subproblems from subproblem 2 by adding the constraint $x_1 \leq 2$ and the constraint $x_1 \geq 3$. The two resulting subproblems are

Subproblem 4   **Subproblem 5**

$$\begin{array}{ll} \max & 2x_1 + x_2 \qquad (3.16) \\ \text{subject to} & x_1 + 4x_2 \leq 9, \\ & 8x_1 + 2x_2 \leq 19 \\ & x_2 \leq 1 \\ & x_1 \leq 2 \\ \text{with} & x_1, x_2 \in \mathbb{N}_0 \text{ and } x_1, x_2 \geq 0. \end{array}$$
$$\begin{array}{ll} \max & 2x_1 + x_2 \qquad (3.17) \\ \text{subject to} & x_1 + 4x_2 \leq 9, \\ & 8x_1 + 2x_2 \leq 19 \\ & x_2 \leq 1 \\ & x_1 \geq 3 \\ \text{with} & x_1, x_2 \in \mathbb{N}_0 \text{ and } x_1, x_2 \geq 0. \end{array}$$

The relaxed subproblem 5 is infeasible and no optimal solution exists. Therefore, this problem is killed and excluded from further consideration. Subproblem 4 is shown in Fig. 3.17. The optimal solution of the relaxed problem is $x^* = (2, 1)$ with $f(x^*) = 5$. As this solution is integral and there are no other active subproblems, we terminate and return $x^* = (2, 1)$ as the optimal solution for the original ILP.

Figure 3.18 shows the search tree describing the search process. Node 1 is the original ILP. Node 2 (subproblem 2) is killed because we found a better integral solution (node 4) and node 5 (subproblem 5) is killed because there is no feasible solution for the relaxed problem. Tree search stops with node 4 as it is the last active node.

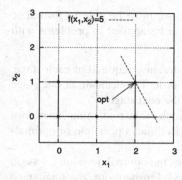

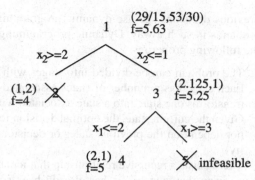

Fig. 3.17 Subproblem 4 created from subproblem 3 by adding the constraint $x_1 \leq 2$

Fig. 3.18 Decision tree when using branch-and-bound for solving the ILP from (3.13)

3.3.4 Dynamic Programming

In branch-and-bound approaches, we start with the original ILP and subsequently create subproblems by introducing additional constraints. *Dynamic programming* is also an exhaustive search method that intelligently enumerates all solutions of a combinatorial optimization problem. It can be applied to problems where the process of finding an optimal solution can be decomposed into separate steps. The idea is to start with the last decision and to work backwards to the earlier ones. To do this, we need a recurrence relation that takes us backwards from a step to the previous step. By working backwards step by step, we find a sequence of optimal decisions and solve the underlying combinatorial optimization problem.

To be able to apply dynamic programming approaches, we have to formulate the problem as a *multistage process*. A multistage process is a process that can be separated into a number of sequential steps (also called stages). The conditions of the process at a given stage are called *states* and a sequence of decisions is called a *policy*. In each stage, we have to make a decision which affects the transition from the current state to a state associated with the next stage.

A multistage decision process is finite if there are only a finite number of stages in the process and a finite number of states associated with each stage. If we define returns (costs or benefits) for each decision, we get an optimization problem. Then, the goal is to find an optimal policy that results in the best total return. In general, we can decide between deterministic and stochastic dynamic programming. In deterministic dynamic programming, given a state and a decision, both the immediate return and next state are known. If one of these is probabilistic, then we have a stochastic dynamic program.

Dynamic programming was proposed by Bellman as an approach to solve multistage decision processes (Bellman, 1952, 1953). It is based on Bellman's *principle of optimality* which says that an optimal policy for a multistage process has the property that no matter what the previous decisions have been, the remaining decisions must constitute an optimal policy with regard to the state resulting from those previous decisions. To use dynamic programming, we must be able to formulate problems in such a way. Dynamic programming can be applied to problems with the following properties:

- The problem can be divided into stages with a decision required at each stage. Each stage has a number of states associated with it. The decision at one stage transforms one state into a state associated with the next stage.
- Given the current state, the optimal decision for each of the remaining states does not depend on the previous states or decisions (Bellman's principle of optimality).
- There exists a recursive relationship that identifies the optimal decision for stage j, given that stage $j + 1$ has already been solved. Furthermore, the final stage must be solvable.

The first item describes general properties of a multistage process. The second item demands that Bellman's optimality principle must hold, and item three requires a recursive relationship between decisions at different stages. Therefore, the main challenge of applying dynamic programming is to define stages and states such that Bellman's optimality principle holds and we can define a recursive relationship. If we can model a problem in such a way, the recursive relationship often makes finding the optimal solution easy.

When solving a problem modeled as a dynamic program, we start with the last stage of a multistage process and determine for each possible state the best policy for leaving that state and completing the process, assuming that all previous stages have been completed. Then, we move backwards through the process and recursively determine for each state the best policy for leaving that state and completing the process, assuming that all previous stages have been completed and using the results already obtained for the succeeding stages.

Formulating problems as dynamic programs is only possible for selected problems. In the literature, a variety of different formulations have been proposed (for an overview see Bellman (2003) or Denardo (2003)). In the following paragraphs, we illustrate how to formulate a special variant of a combinatorial optimization problem as a dynamic program. The problem has the following structure:

$$\max \quad z = f_1(x_1) + f_2(x_2) + \ldots + f_n(x_n) = \sum_{i=1}^{n} f_i(x_i) \qquad (3.18)$$

$$\text{subject to} \quad x_1 + x_2 + \ldots + x_n \leq b,$$

$$\text{with} \quad x \in \mathbb{N}_0, \, b \geq 0$$

where $f_i(x_i)$ are (non-linear) functions of single variables. All problems, linear or non-linear, that have this form can be formulated as a multistage process with n stages. In each stage (except the first one), we have $b + 1$ different states u, where $u \in \{0, \ldots, b\}$. The value of state u_j chosen in stage j is determined by the integral decision variable $x_j \in \{0, 1, \ldots, b\}$. For $j > 0$, the number of states in stage j equals the number of possible values of the jth decision variable, $|u_j| = |x_j|$. In the first stage, we have only a single state b. In stage $j \in \{1, \ldots, n\}$, we specify possible values for the decision variable x_j that determine the state u_j and contribute with $f_j(x_j)$ to the overall return.

When solving this multistage problem using dynamic programming, we must formulate a recursion on how the optimum return from completing the process depends on beginning at state u in stage j. This recursion is problem-dependent and different for different types of multistage processes. In general, for formulating a recursion we need a termination criterion and a recursion step. The termination criterion determines how in the last stage the return depends on the state u. The recursion step relates the return of a state in stage j to the return of stage $j+1$. For problems in the form (3.18), we can calculate in the last stage n the optimum return $m_n(u)$ of state u as

$$m_n(u) = \max_{0 \leq x_n \leq u} (f_n(x_n)), \tag{3.19}$$

where $u = 0, 1, \ldots, b$. u is the state variable, whose values specify the states. We also need a recursion step for calculating the optimum return $m_j(u)$ from completing the process beginning at stage j in state u:

$$m_j(u) = \max_{0 \leq x_j \leq u} (f_j(x_j) + m_{j+1}(u - x_j)), \tag{3.20}$$

where $j = n-1, \ldots, 1$. We can solve the problem by starting with stage n and recursively determining $m_j(u)$. The optimal solution x^* has an objective value of $m_1(b)$. We want to illustrate the functionality of the approach for an example (non-linear problem) which is defined as

$$\begin{aligned} \max \quad & z = 1.2\sqrt{x_1} + \log(1 + 20x_2) + \sqrt{x_3} \tag{3.21} \\ \text{subject to} \quad & x_1 + x_2 + x_3 \leq 3, \\ \text{with} \quad & x_i \in \mathbb{N}_0, \end{aligned}$$

where $f_1(x_1) = 1.2\sqrt{x_1}$, $f_2(x_2) = \log(1 + 20x_2)$, and $f_3(x_3) = \sqrt{x_3}$. This is a non-linear multistage problem with $n = 3$ stages and a maximum of $b + 1 = 4$ different states at each stage. When solving the problem, we start with the last stage of the process, assuming that all previous steps (stages 1 and 2) have been completed. Therefore, we have four possible states and x_3 can be either 0, 1, 2, or 3. Using (3.19), we get

$$\begin{aligned} m_3(3) &= \max(f_3(0), f_3(1), f_3(2), f_3(3)) = \max(\sqrt{0}, \sqrt{1}, \sqrt{2}, \sqrt{3}) = \sqrt{3}, \\ m_3(2) &= \max(f_3(0), f_3(1), f_3(2)) = \max(\sqrt{0}, \sqrt{1}, \sqrt{2}) = \sqrt{2}, \\ m_3(1) &= \max(f_3(0), f_3(1)) = \max(\sqrt{0}, \sqrt{1}) = 1, \\ m_3(0) &= \max(f_3(0)) = \max(\sqrt{0}) = 0. \end{aligned}$$

We continue with stage 2 and use (3.20) to calculate the maximum returns for the four possible states.

$$m_2(3) = \max(f_2(0) + m_3(3-0), f_2(1) + m_3(3-1), f_2(2) + m_3(3-2), f_2(3)$$
$$+ m_3(3-3)) = \max(\log(1) + \sqrt{3}, \log(21) + \sqrt{2}, \log(41) + 1, \log(61) + 0)$$
$$= \log(21) + \sqrt{2}$$
$$m_2(2) = \max(f_2(0) + m_3(2-0), f_2(1) + m_3(2-1), f_2(2) + m_3(2-2))$$
$$= \max(\log(1) + \sqrt{2}, \log(21) + 1, \log(41) + 0)$$
$$= \log(21) + 1$$
$$m_2(1) = \max(f_2(0) + m_3(1-0), f_2(1) + m_3(1-1)) = \max(\log(1) + 1, \log(21) + 0)$$
$$= \log(21)$$
$$m_2(0) = \max(f_2(0) + m_3(0-0)) = 0$$

After completing state 2, we turn to stage 1. There is only one state associated with this stage, $u = 3$. We get

$$m_1(3) = \max(f_1(0) + m_2(3-0), f_1(1) + m_2(3-1), f_1(2) + m_2(3-2), f_1(3)$$
$$+ m_2(3-3)) = \max(1.2\sqrt{0} + \log(21) + \sqrt{2}, 1.2\sqrt{1}$$
$$+ \log(21) + 1, 1.2\sqrt{2} + \log(21), 1.2\sqrt{3} + 0$$
$$= 2.2 + \log(21)$$

Thus, the optimal solution $x^* = (1,1,1)$ has an objective value of $f(x^*) = 2.2 + \log(21) \approx 3.5$.

As a second example, we want to formulate the TSP as a dynamic program (Held and Karp, 1962). There are n cities and we start the tour from city o. Formulating this problem as a multistage problem, we have $n-1$ different stages and in each stage we have to decide on the next city we want to visit. We only need $n-1$ stages as the start city o is randomly fixed and, thus, need not be considered for the recursion. A state u is represented by a pair (i, S), where S is the set of j cities that are already visited and i is the city visited in stage j ($i \in S$). Assuming that the cities are labeled from 1 to n we can define the termination criterion for the recursion as

$$m_{n-1}(i, S) = d(i, o), \tag{3.22}$$

where o is a randomly chosen start city, $S = \{\{1,\ldots,n\} - \{o\}\}$ and $i \in \{\{1,\ldots,n\} - \{o\}\}$. $d(i, j)$ denotes the distance between two cities i and j. For other stages, the recursion is

$$m_j(i, S) = \min_{l \neq o,\, l \notin S} (d(i, l) + m_{j+1}(l, S \cup l)). \tag{3.23}$$

We have $n-1$ different stages and at each stage j we have to consider $n-1$ different cities; all sets are of size j. The time and space complexity of this dynamic programming approach is $O(n2^n)$ (Held and Karp, 1962). Although the effort still increases exponentially, it is much lower than complete enumeration which is $O((n-1)!)$.

We want to illustrate the functionality of this approach for the example TSP with four nodes from Sect. 3.3.2 (p. 65). Assuming that we start with city a ($o = a$), we can use (3.22) to calculate the optimal return $m_3(i,S)$ from completing the process at stage 3;

$$m_3(b,\{b,c,d\})=d(b,a)=6$$
$$m_3(c,\{b,c,d\})=d(c,a)=8$$
$$m_3(d,\{b,c,d\})=d(d,a)=1$$

In stage 2, there are six states and their optimal return $m_2(i,S)$ can be calculated according to (3.23) as:

$$m_2(b,\{b,c\}) = (d(b,d)+m_3(d,\{b,c\}\cup\{d\})) = d(b,d)+m_3(d,\{b,c,d\})$$
$$= 1+1 = 2,$$
$$m_2(b,\{b,d\}) = (d(b,c)+m_3(c,\{b,d\}\cup\{c\})) = d(b,c)+m_3(c,\{b,c,d\})$$
$$= 2+8 = 10,$$
$$m_2(c,\{b,c\}) = (d(c,d)+m_3(d,\{b,c\}\cup\{d\})) = d(c,d)+m_3(d,\{b,c,d\})$$
$$= 2+1 = 3,$$
$$m_2(c,\{c,d\}) = (d(c,b)+m_3(b,\{c,d\}\cup\{b\})) = d(c,b)+m_3(b,\{b,c,d\})$$
$$= 2+6 = 8,$$
$$m_2(d,\{c,d\}) = (d(d,b)+m_3(b,\{c,d\}\cup\{b\})) = d(d,b)+m_3(b,\{b,c,d\})$$
$$= 1+6 = 7,$$
$$m_2(d,\{b,d\}) = (d(d,c)+m_3(c,\{b,d\}\cup\{c\})) = d(d,c)+m_3(c,\{b,c,d\})$$
$$= 2+8 = 10.$$

In stage one, we have three possible states and get

$$m_1(b,\{b\}) = \min(d(b,c)+m_2(c,\{b\}\cup\{c\}),d(b,d)+m_2(d,\{b\}\cup\{d\}))$$
$$= \min(2+3;1+10) = 5,$$
$$m_1(c,\{c\}) = \min(d(c,b)+m_2(b,\{c\}\cup\{b\}),d(c,d)+m_2(d,\{c\}\cup\{d\}))$$
$$= \min(2+2;2+7) = 4,$$
$$m_1(d,\{d\}) = \min(d(d,b)+m_2(b,\{d\}\cup\{b\}),d(b,c)+m_2(c,\{d\}\cup\{c\}))$$
$$= \min(1+10;2+8) = 10.$$

Although the recursion is completed, we have not yet considered traveling from city a to the first city. We can either introduce an additional step in the recursion or calculate modified optimal returns as

$$m_1'(b,\{b\}) = m_1(b,\{b\})+d(a,b) = 5+6 = 11,$$
$$m_1'(c,\{c\}) = m_1(c,\{c\})+d(a,c) = 4+8 = 12,$$
$$m_1'(d,\{d\}) = m_1(d,\{d\})+d(a,d) = 10+1 = 11.$$

We have found two optimal solutions (*abcda* and *adcba*) with tour cost 11.

This section illustrated the functionality of dynamic programming which can be used to solve ILPs as well as non-linear problems if they can be formulated as a multistage process. A problem is solved by decomposing the solution process into single steps and defining a recursion that relates the different steps. Dynamic programming is an exact approach that enumerates the search space and intelligently discards inferior solutions. The time and space complexity of dynamic programming depends on the problem considered but is usually much lower than complete enumeration. For further details on dynamic programming, we refer to Bellman (2003) and Denardo (2003). A nice summary including additional exercises can be found in Bronson and Naadimuthu (1997, Chap. 19).

3.3.5 Cutting Plane Methods

Cutting plane methods are approaches for solving ILPs that can either be used alone or in combination with other techniques like branch-and-bound. Cutting plane methods are based on the idea to add additional constraints (*cutting planes*) to a problem such that infeasible fractional solutions (including the optimal, but fractional, solution) are removed but all integer solutions remain feasible. When performing such iterative cuts by adding constraints to an ILP, the original set of constraints is replaced by alternative constraints that are closer to the feasible integral solutions and exclude fractional solutions.

In Sect. 3.2.2, we discussed how a convex polyhedron, which represents the feasible search space of an LP, can be described as the intersection of half-spaces (Weyl, 1935), where each half-space is defined by a constraint. When using the Simplex method for solving LPs, we move on the convex hull defined by the half-spaces and iteratively examine corner points until we find the optimal solution. By analogy to LPs, the feasible search space of an ILP can also be described by using a convex hull where the corner points are integral solutions. Thus, if we can find a set of linear inequalities that completely defines the feasible search space of an ILP such that the corner points are integral, then we can solve an ILP using the Simplex method. The goal of cutting plane methods is to obtain that integral convex hull from the fractional convex hull of the underlying linear program by introducing additional constraints that cut away fractional corner points.

In general, there are two different ways to generate cuts. First, we can generate cuts based on the structure of the problem. Such cuts are problem-specific and have to be generated separately for each problem. However, once good cuts are found, these cuts provide very efficient solution techniques. An example is problem-specific cuts for the TSP (Dantzig et al, 1954) which inspired *branch-and-cut algorithms* (Grötschel et al, 1984; Padberg and Rinaldi, 1991). Branch-and-cut algorithms are problem-specific methods that combine cutting plane methods with a branch-and-bound algorithm. Branch-and-cut methods systematically attempt to obtain stronger (in the sense that the optimal solution is less fractional) LP relaxations at every node of the search tree by introducing additional cutting planes.

The second way is to iteratively solve the relaxed ILP using the Simplex method and generate cuts that remove the optimal fractional solution from the Simplex. Such approaches are not problem-specific and can be applied to any ILP. The first successful approach following this idea was presented by Gomory (1958) who developed a *cutting plane method* for ILPs which obtains the integral convex hull after applying a sequence of cuts (linear constraints) on the fractional convex hull (Gomory, 1960, 1963). Chvátal (1973) showed that this procedure always terminates in a finite number of steps and converges to an optimal solution obtaining the integral convex hull. However, the number of steps can be very large due to the fact that these algebraically-derived cuts are often weak in the sense that the area that is cut from the Simplex is small. Furthermore, the performance of cutting plane algorithms is limited as the minimal number of inequalities that are necessary to describe the integral convex hull increases exponentially with the number of decision variables. However, usually we do not need to correctly describe the complete convex hull but only need a partial, accurate description of the convex hull in the neighborhood of the optimal solution. Therefore, in practice, cutting plane approaches show good performance for a variety of different combinatorial optimization problems and many implementations for solving ILPs use cutting plane methods. Usually, cutting plane methods for ILPs iteratively perform three steps:

1. We drop the integral constraint and solve the relaxed ILP obtaining the optimal solution x^*.
2. If the relaxed ILP is unbounded or infeasible, so is the original ILP. If x^* is integral, the problem is solved and we can stop.
3. If not, we add a linear inequality constraint to the problem such that all integral solutions remain feasible and x^* becomes infeasible. We continue with step 1.

The main difficulty of cutting plane methods is to generate a proper constraint. The following paragraphs describe how the *Gomory cutting plane algorithm* (Gomory, 1958) generates cuts for ILPs in standard form:

$$\min \quad c^T x \tag{3.24}$$
$$\text{subject to} \quad Ax = b$$
$$x_i \in \mathbb{N}_0$$

In this formulation, the vector x contains both the original set of decision variables as well as the slack variables (see Sect. 3.2.2). b is assumed to be integral.

We can solve the relaxed ILP using the Simplex method and obtain the optimal solution x^*. Let m be the number of constraints, which is equal to the number of rows of A. We assume that the variables $\{x_1, \ldots, x_m\}$ are basis variables and the variables $\{x_{m+1}, \ldots, x_n\}$ are non-basis variables. Then, we can reformulate problem (3.24) as

$$I x_b + \bar{A} x_f = \bar{b} \tag{3.25}$$

where I is the identity matrix, $x_b = (x_1, \ldots, x_m)^T$ is the vector of basic variables, $\bar{A}$ is a matrix that depends on b and A, $x_f = (x_{m+1}, \ldots, x_n)^T$ is the vector of non-basic variables, and $\bar{b}$ is a vector of size n. We have n different constraints and each row of this problem formulation corresponds to

$$x_i + \sum_{j \in \{m+1,\ldots,n\}} \bar{a}_{ij} x_j = \bar{b}_i \qquad (3.26)$$

where $i \in \{1, \ldots, m\}$. As we know that all $x_j \geq 0$, we can rewrite (3.26) as

$$x_i + \sum_{j \in \{m+1,\ldots,n\}} \lfloor \bar{a}_{ij} \rfloor x_j \leq x_i + \sum_{j \in \{m+1,\ldots,n\}} \bar{a}_{ij} x_j = \bar{b}_i.$$

If the constraints are in form (3.25), then $x_i^* = b_i$ as we can set the slack variables to $x_j = 0$. Therefore, to get an optimal integral solution, the b_i must be integral. Consequently, we are now able to formulate a *Gomory cut* as

$$x_i + \sum_{j \in N} \lfloor \bar{a}_{ij} \rfloor x_j \leq \lfloor \bar{b}_i \rfloor. \qquad (3.27)$$

Substituting x_i by using (3.26) we can reformulate this cut and get

$$\sum_{j \in N} (\bar{a}_{ij} - \lfloor \bar{a}_{ij} \rfloor) x_j \geq \bar{b}_i - \lfloor \bar{b}_i \rfloor. \qquad (3.28)$$

In this formulation of the cut, no basis variables are included. If the optimal solution x^* is fractional, it does not satisfy constraint (3.27) or (3.28). However, no feasible integral solutions are removed from the search space using this cut.

We want to use the problem defined in (3.13) to study Gomory cuts. After introducing the slack variables x_3 and x_4, we can formulate the constraints for this maximization problem as

$$\begin{aligned} x_1 \;\; +4x_2 \;\; +x_3 \qquad\quad &= 9 \\ 8x_1 \;\; +2x_2 \qquad\quad +x_4 &= 19 \end{aligned} \qquad (3.29)$$

Transforming the constraints into form (3.25) results in:

$$\begin{aligned} x_1 \qquad -\tfrac{1}{15}x_3 + \tfrac{2}{15}x_4 &= \tfrac{29}{15} \\ x_2 + \tfrac{4}{15}x_3 - \tfrac{1}{30}x_4 &= \tfrac{53}{30} \end{aligned} \qquad (3.30)$$

As already seen in Sect. 3.3.3, the optimal solution of the relaxed problem is $x^* = \left(\tfrac{29}{15}, \tfrac{53}{30}\right)$.

Now we want to find an additional constraint (Gomory cut) for this problem. As both basis variables x_1^* and x_2^* are fractional, we have two possible cuts. According to (3.27) we get

$$\text{cut 1:} \quad x_1 + \lfloor -\tfrac{1}{15} \rfloor x_3 + \lfloor \tfrac{2}{15} \rfloor x_4 \leq \lfloor \tfrac{29}{15} \rfloor \qquad (3.31)$$

$$\text{cut 2:} \quad x_2 + \lfloor \tfrac{4}{15} \rfloor x_3 + \lfloor -\tfrac{1}{30} \rfloor x_4 \leq \lfloor \tfrac{53}{30} \rfloor \qquad (3.32)$$

We choose cut 1 and obtain the constraint $x_1 - x_3 \leq 1$. Using (3.29), cut 1 can be formulated as $2x_1 + 4x_2 \leq 10$. Figure 3.19(a) shows the resulting problem with the new cut. The optimal solution x^* of the relaxed problem is $x^* = (2, \frac{3}{2})$ with $f(x^*) = 5.5$. As the optimal solution is not yet integral, we want to find another cut. We add cut 1 to the set of constraints (3.30) and obtain

$$
\begin{aligned}
x_1 \quad -\tfrac{1}{15}x_3 + \tfrac{2}{15}x_4 \quad &= \tfrac{29}{15} \\
x_2 + \tfrac{4}{15}x_3 - \tfrac{1}{30}x_4 \quad &= \tfrac{53}{30} \\
x_1 \quad -x_3 \quad\quad +x_5 &= 1
\end{aligned} \tag{3.33}
$$

Again we have to bring the constraints into form (3.25). This transformation yields

$$
\begin{aligned}
x_1 \quad +\tfrac{1}{7}x_4 - \tfrac{1}{14}x_5 &= 2 \\
x_2 \quad -\tfrac{1}{14}x_4 + \tfrac{2}{7}x_5 &= \tfrac{3}{2} \\
x_3 +\tfrac{1}{7}x_4 - \tfrac{15}{14}x_5 &= 1
\end{aligned} \tag{3.34}
$$

x_1, x_2, and x_3 are the basis variables. As only x_2^* is fractional, there is only one possible cut $x_2 - x_4 \leq 1$. Using (3.34), we can reformulate this cut as $8x_1 + 3x_2 \leq 20$. Figure 3.19(b) shows the cut and the optimum solution of the relaxed ILP $x^* = (\frac{25}{13}, \frac{20}{13})$ with $f(x^*) = \frac{70}{13}$.

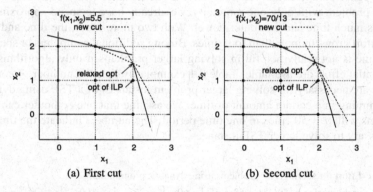

Fig. 3.19 Two Gomory cuts for the problem defined in (3.13)

Here we will stop generating more Gomory cuts for the example problem and leave finding additional cuts to the reader. The example illustrates the relevant properties of cutting plane methods. Adding cuts to the problem reduces the size of the feasible search space and removes fractional solutions. However, such improvements are usually small and often we need a large number of cuts to find an optimal integral solution.

3.4 Heuristic Optimization Methods

In the previous sections, we presented a variety of standard optimization methods for combinatorial optimization problems. All previously presented methods are exact methods as they guarantee finding the optimal solution. Usually, exact optimization methods are efficient problem solvers if they solve a problem with polynomial effort. A nice example of efficient methods for linear, continuous optimization problems are interior point methods as their worst-case running time increases polynomially with the problem size.

However, there are many interesting optimization problems where no polynomial optimization methods are known. For example, many combinatorial optimization problems like the traveling salesman problem, routing or assignment problems are NP-hard (Garey and Johnson, 1979) and so the effort of exact optimization methods applied to such problems increases exponentially with the problem size. Therefore, all exact optimization methods that are applied to NP-hard problems are only useful for small problem instances; larger problems become intractable as the effort for solving the problem increases exponentially with n.

We want to illustrate the problem of exponential complexity for the TSP. When using an exact method for this problem, its effort increases exponentially with the number n of cities. For example, when solving this problem with dynamic programming (see the example in Sect. 3.3.4), space and time complexity is $O(n2^n)$. If we have bought a computer that is able to solve a TSP instance with n cities and want to solve a problem with only one more city, we already have to spend approximately twice as much time and computer power. With two more cities, the time and computer effort grows to more than four times. Buying additional computers or spending more time is not really helpful in solving larger problems if only algorithms with exponential effort are available. Table 3.1 exemplarily lists the number of computers that are necessary for solving larger problem instances of a TSP using dynamic programming in a certain amount of time. We assume that one computer can solve problems with $n = 50$ cities in one time period. The numbers indicate the time that is necessary to solve larger TSP instances.

Table 3.1 Effort for solving TSP problems using dynamic programming

problem size n	50	51	52	55	60	70	100	200
necessary effort	1	2.04	4.16	35.2	1,228.8	1.47×10^6	2.25×10^{15}	5.7×10^{45}

As many interesting problems are intractable and only small problem instances can be solved exactly, methods have been developed that do not guarantee finding an optimal solutions but whose running time is lower than the running time of exact optimization methods. Usually, such methods are problem-specific. We want to denote optimization methods that do not guarantee to find the optimal solution and which use problem-specific (heuristic) information about the problem as *heuristic optimization methods*. The necessity to use problem-specific information in heuristic optimization methods is a result of the *no-free-lunch theorem*. The theorem says that no *black-box optimization* is possible but efficient optimization methods need to be problem-specific. This is in contrast to the goal of black-box optimization which aims at building a general-purpose solver in the sense that we just hand over our problem to the solver and the method returns, in reasonable time, an optimal or at least high-quality solution. The no-free-lunch theorem says that this is not possible; for example, on problems with a trivial topology, where no useful metric can be defined (see Sect. 2.3.1, p. 16), no optimization techniques can perform better than others but all show on average the same performance. We can overcome this limitation if problems have metric search spaces and optimization methods are problem-specific, exploiting meaningful and problem-specific structures of a search space.

Therefore, usually heuristic optimization methods are problem-specific (in contrast to, for example, random search or simple enumeration methods like breath and depth search). However, when designing heuristic optimization methods, we must ensure that knowledge about the problem which is used by the heuristic optimization method also holds for the problem that should be solved. Using the "right" heuristic for a "wrong" problem which does not show the characteristics that are exploited by the heuristic optimization method leads to low solution quality.

The concept of heuristic optimization methods is old and first appeared in the literature in the early 1940s (Polya, 1945). By the end of the early 1960s, it was a common concept in computer science. Heuristic methods were seen as *'provisional and plausible procedures whose purpose is to discover the solution of a particular problem at hand'* (Gelernter, 1963, p. 192). At the early years, a heuristic was a *'rule of thumb'* for solving a problem (Simon and Newell, 1958). Heuristics are problem-specific and exploit known rules of thumb, tricks or simplifications to obtain a high-quality solution for the problem. Heuristics do not guarantee finding an optimal solution but are usually faster than approaches returning an optimum. Therefore, heuristic optimization does not aim at finding an optimal solution but at developing optimum solution procedures (Kuehn and Hamburger, 1963). Heuristics do not guarantee optimality (this would be a contradiction of the word "heuristics") but are instead designed to be fast and problem-specific. Based on Foulds (1983), Silver (2004, p. 936) defined a heuristic as *'a method which, on the basis of experience or judgement, seems likely to yield a reasonable solution to a problem, but which cannot be guaranteed to produce the mathematically optimal solution'*. For an overview of the development of the term heuristics, we refer to Romanycia and Pelletier (1985) or Silver (2004). For the purpose of this book, we want to distinguish three types of heuristic optimization methods:

- *heuristics*,
- *approximation algorithms*, and
- *modern heuristics*.

There are two different types of heuristics: *construction heuristics* that construct one solution from scratch by performing iterative construction steps and *improvement heuristics* that start with a complete solution and iteratively apply problem-specific search operators to search through the search space. Usually, construction heuristics use problem-specific rules for the construction of a solution and improvement heuristics apply problem-specific search operators and search strategies.

Approximation algorithms are heuristic optimization methods that return an approximate solution with guaranteed solution quality. Therefore, for approximation algorithms we are able to provide a bound on the quality of the returned solution. The main difference between heuristics and approximation algorithms is the existence of a bound on the solution quality. If we are able to formulate a bound, a heuristic "becomes" an approximation algorithm.

Finally, we denote improvement heuristics that use a general, problem-invariant, and widely applicable search strategy as *modern heuristics* (Polya, 1945; Romanycia and Pelletier, 1985; Reeves, 1993; Rayward-Smith et al, 1996; Rayward-Smith, 1998; Michalewicz and Fogel, 2004). Such methods are denoted as modern heuristics since they define a strategy for searching through the search space on a meta-level. Therefore, they are, especially in the OR community, also often denoted as *metaheuristics* (Glover, 1986). Modern heuristics often imitate the functionality of search strategies observed in other domains (for example in biology, nature, or physics). Characteristic properties of modern heuristics are that the same search concept can successfully be applied to a relatively wide range of different problems and that *intensification* and *diversification* are alternately used in the search. During intensification, modern heuristics focus their search on promising areas of the search space and during diversification new areas of the search space are explored.

A precise distinction between improvement heuristics and modern heuristics is difficult as the concepts are closely related and in the literature no consistent naming convention exists. Usually, modern heuristics are defined problem-independently, whereas heuristics are explicitly problem-specific and exploit problem structure. Therefore, the design and application of high-quality heuristics is demanding, since problem-specific properties must be known and actively exploited. In contrast, standard modern heuristics can easily be applied to different problems with little or no modification. There is a trade-off between ease of application and effectiveness. Problem-specific heuristics are more difficult to develop and apply in comparison to modern heuristics. However, if problem-specific knowledge is considered appropriately, they often outperform standard modern heuristics. Later in this book, we discuss what types of problem characteristics are exploited by modern heuristics (e.g. locality or decomposability) and how the performance of modern heuristics can also be improved by problem-specific knowledge.

The following sections provide a brief overview of the different types of heuristic optimization methods. We start with heuristics and continue with approximation algorithms (Sect. 3.4.2) and modern heuristics (Sect. 3.4.3). The chapter ends with the no-free-lunch theorem.

3.4.1 Heuristics

Heuristics are optimization methods that try to exploit problem-specific knowledge and for which we have no guarantee that they find the optimal solution. Heuristics are often defined based on knowledge about high-quality solutions and also incorporate rules of thumb (Simon and Newell, 1958; Pearl, 1984). The design of efficient heuristics is demanding as we must have some knowledge about the structure of the problem to be solved and we must have some idea of what distinguishes high-quality solutions from low-quality solutions.

When dealing with optimization methods, we can observe a general trade-off between effectiveness and application range. Usually, the more problems we want to solve with a particular method, the lower is the method's performance. Heuristics are usually designed for a particular problem and try to exploit problem-specific knowledge. Therefore, by increasing problem specificity and narrowing the application scope we can often increase their performance. In the extreme case, we have a highly-effective heuristic that yields high-quality solutions for one particular type of problem but fails for all other related problems with a slightly different structure.

Although there is early work that describes how heuristics can be used for problem solving (Polya, 1948; Simon and Newell, 1958; Tonge, 1961; Wiest, 1966; Newell, 1969; Heroelen, 1972), until the 1970s mainly exact optimization algorithms were developed. The situation changed at the beginning of the 1970s as it became clear that many problems of practical relevance are NP-complete and (if $P \neq NP$) exact methods cannot solve such problems in polynomial time (Cook, 1971; Karp, 1972, see also Sect. 2.4.1). Since then, a substantial number of researchers turned from trying to develop exact algorithms to devising approximation algorithms (see Sect. 3.4.2). Approximation algorithms are polynomial time heuristics that provide some sort of guarantee on the quality of the found solution.

Heuristics are especially popular in operations research and artificial intelligence (Schwefel, 1965; Holland, 1975; Goldberg, 1989c; Koza, 1992; Mitchell, 1996; Bäck et al, 1997; Glover and Kochenberger, 2003; Reeves and Rowe, 2003; Silver, 2004; Gendreau and Potvin, 2005). In these areas, there has been only limited interest in developing bounds on the solution quality but the performance of heuristics is mainly evaluated using experimental comparisons between different methods (Bartz-Beielstein, 2006). Furthermore, heuristics are often viewed as a simplified form of modern heuristics. We want to distinguish between two types of heuristics:

- construction heuristics and
- improvement heuristics.

Construction heuristics (which are also known as single-pass heuristics) build a solution from scratch in a step-wise creation process where in each step parts of the solution are fixed. For combinatorial optimization problems, in each iteration often only one decision variable is determined. Construction heuristics do not attempt to improve the solution once it is constructed but stop after obtaining a complete solution. Solving a TSP by using greedy search as we have presented in Sect. 3.3.2 (p. 66 and Fig. 3.12) illustrates the functionality of a construction heuristic where in each iteration one city is added to a tour. In this example, the heuristic iteratively chooses a city that minimizes the estimated length of a tour back to the starting city.

Improvement heuristics start with a complete solution and iteratively try to improve the solution. In contrast to modern heuristics where improvement steps alternate with diversification steps (diversification steps usually lead to solutions with a worse objective value), improvement heuristics use no explicit diversification steps. Often, improvement heuristics perform only improvement steps and stop if the current solution can not be improved any more and a local optimum has been reached. By performing iterative improvement steps, we define a metric on the search space (see Sect. 2.3.2) as all possible solutions that can be reached in one improvement step are neighboring solutions. For the design of appropriate neighborhoods of improvement heuristics, we can use the same criteria as for modern heuristics (see Chap. 4.2). An example of an improvement heuristic is given in Fig. 3.14, p. 68. An improvement heuristic for the TSP starts with a complete solution and iteratively examines all neighboring solutions. It stops at a local optimum, where all neighbors have a worse objective value.

Both types of heuristics, improvement heuristics as well as construction heuristics, are often based on greedy search. Greedy search (see also Sect. 3.3.2) is an iterative search approach that uses a heuristic function h to guide the search process. h estimates the minimal distance to an optimal solution. In each search step, greedy search chooses the solution where the heuristic function becomes minimal, i.e. the improvement of the solution is maximal. A representative example of greedy search is best-first search (Sect. 3.3.2).

As heuristics are problem-specific, only representative examples can be given. In the following paragraphs, we present selected heuristics for the TSP illustrating the functionality of heuristics. In the TSP, the goal is to find a tour of n cities with minimal tour length. We can distinguish between construction and improvement heuristics. Construction heuristics start with an empty tour and iteratively add cities to the tour until it is completed. Improvement heuristics start with a complete tour and perform iterative improvements. Representative examples of construction heuristics for the TSP are:

1. **Nearest neighbor heuristic:** This heuristic (Rosenkrantz et al, 1977) chooses a starting city at random and iteratively constructs a tour by going from the current city to the nearest city that is not yet included in the tour. After adding all cities, the tour is completed by connecting the last city with the starting city. This heuristic does not perform well in practice, although for non-negative distances that satisfy the triangle inequality $(d(x,y) \leq d(x,z) + d(z,y))$ the ratio between the length $l(T)$ of the tour T found by the heuristic and the length of the optimal tour T_{opt} never exceeds $(\log_2 n)/2$. Therefore, we have an upper bound on the solution quality $l(T)/l(T_{opt}) \leq (\log_2 n)/2$.

2. **Nearest insertion heuristic:** We start with a tour between two cities and iteratively add the city whose distance to any city in the tour is minimal. The city is added to the tour in such a way that the increase in the tour length is minimal. A city can be either added at the endpoints of the tour or by removing one

edge and inserting the new city between the two new ends. We have a worst-case performance ratio of $l(T)/l(T_{opt}) \leq 2$.

3. **Cheapest insertion heuristic:** This heuristic works analogously to the nearest insertion heuristic but chooses the city to insert as the one which increases the cost of the tour the least. The bound on the solution quality is $l(T)/l(T_{opt}) \leq \log_2 n$.

4. **Furthest insertion heuristic:** This heuristic starts with a tour that visits the two cities which are furthest apart. The next city to be inserted is the one that increases the length of the current tour the most when this city is inserted in the best position on the current tour. The idea behind this heuristic is to first insert cities that are far apart. The worst-case performance ratio is $l(T)/l(T_{opt}) \leq \log_2 n$. Although the performance guarantee is the same as for the cheapest insertion heuristic, in practice furthest insertion consistently outperforms the other construction heuristics.

In the literature, there are a variety of other construction heuristics for TSPs. Another relevant example of construction heuristics is the Christofides heuristic which is discussed in Sect. 3.4.2 as a representative example of an approximation algorithm.

Furthermore, there are also examples of representative and effective improvement heuristics for the TSP.

1. **Two-opt heuristic**: This heuristic iteratively removes all possible pairs of non-adjacent edges in a tour and connects the two resulting unconnected sub-tours such that the tour length is minimal. If the distances between all cities are Euclidean (the cities are on a two-dimensional plane), two-opt returns a non-crossing tour.

2. **k-opt heuristic:** k-opt (Lin, 1965) is the generalization of 2-opt. The idea is to examine some or all $\binom{n}{k}$ k-subsets of edges in a tour. For each possible subset S, we test whether an improved tour can be constructed by replacing the edges in S with k new edges. Such a replacement is called a *k-switch*. A tour where the application of a k-switch does not reduce the tour length, is denoted as a *k-optimum*. A k-opt heuristic iteratively performs k-switches until a k-optimum is obtained. Although the effort for k-opt is high and only 2-opt and 3-opt are practical, the expected number of iterations required by 2-opt is polynomial for Euclidean problem instances and runs in $O(n^{10} \log n)$ (Chandra et al, 1994). For TSPs satisfying the triangle inequality the worst-case performance ratio of 2-opt is $O(4\sqrt{n})$ and for k-opt it is $O(\frac{1}{4}n^{\frac{1}{2k}})$ (Chandra et al, 1994).

3. **Lin-Kernighan heuristic:** This heuristic (Lin and Kernighan, 1973) is an extension of the k-opt heuristic. In contrast to k-opt, the Lin-Kernighan heuristic allows k to vary during the search and does not necessarily use all improvements immediately after they are found. The heuristic starts by removing an edge from a random tour creating a path. Then one end of this path is connected to some internal node creating a cycle with an additional tail. Then, an edge is again removed, thereby yielding a new path. These add/remove operations are repeated as long as the "gain sum" is positive and there still remain unadded/unremoved edges. The "gain sum" is the difference between the sum of the lengths of the

removed edges (except the last) and the sum of the lengths of the added edges. For each path constructed, the cost of the tour obtained by joining its endpoints is also computed and if it is lower than the original tour, the heuristic keeps track of this solution. When a sequence of possible interchanges has been exhausted and an improvement was found during the add/remove operations, the best tour found replaces the original tour and the procedure is repeated. Although it is possible to construct problem instances where the performance of the Lin-Kernighan heuristic is low (Papadimitriou and Steiglitz, 1978), it is a powerful heuristic for TSPs and shows good performance for large real-world problem instances (Helsgaun, 2000). Numerous improvements on the basic strategy have been proposed over the years and state-of-the-art implementations of Lin-Kernighan heuristics are the methods of choice for practitioners.

3.4.2 Approximation Algorithms

We have seen in the previous section that heuristics are an interesting and promising approach to obtain high-quality solutions for intractable problems. Although heuristics provide no guarantee of finding the optimal solution, they often find "good" solutions with polynomial effort. Therefore, heuristics are substituting optimality by tractability.

Approximation algorithms are an attempt to formalize heuristics. Approximation algorithms emerged from the field of theoretical computer science (Johnson, 1974; Hochbaum and Shmoys, 1987; Hochbaum, 1996; Vazirani, 2003) but are also becoming popular in the field of artificial intelligence, especially evolutionary algorithms (Droste et al, 2002; Jansen and Wegener, 2002; Beyer et al, 2002; Neumann and Wegener, 2007; Neumann, 2007). As we know that there are intractable problems (for example NP-complete problems), we are interested in heuristics that have polynomial running time and for which we can develop bounds on the quality of the returned solution. Therefore, approximation algorithms are heuristics for which we can derive a bound on the quality of the returned solution.

The performance of approximation algorithms is measured using the *approximation ratio*

$$\rho(n) \geq \max \left(\frac{f(x^{approx})}{f(x^*)}, \frac{f(x^*)}{f(x^{approx})} \right),$$

where n is the problem size, x^{approx} is the solution returned by an approximation algorithm and x^* is the optimal solution. $\rho(n)$ represents a bound on the solution quality and measures the worst-case performance of an algorithm. This definition of $\rho(n)$ holds for minimization and maximization problems. We say that a heuristic has an approximation ratio of $\rho(n)$ if for any input of size n the objective value $f(x^{approx})$ is within a factor of $\rho(n)$ of the objective value $f(x^*)$. If an algorithm always returns the optimal solution, $\rho(n) = 1$.

As approximation algorithms are heuristics, there exist a large variety of different approximation algorithms for different types of problems. Often, we can observe a trade-off between the quality of an approximation and problem-specificity. Usually, an increasing number of problems where an approximation algorithm can be applied leads to a higher approximation ratio.

Furthermore, there are approximation algorithms where a trade-off exists between computational effort and solution quality. For some NP-complete problems, polynomial-time approximation algorithms are available that can achieve increasingly smaller approximation ratios ($\rho(n) \to 1$) by using more and more computation time. Such *approximation schemes* are approximation algorithms that result, for any fixed $\varepsilon > 0$, in an approximation ratio of $1 + \varepsilon$. Consequently, we want to distinguish three types of approximation algorithms:

- **Fully polynomial-time approximation scheme (FPAS or FPTAS):** An approximation algorithm is an FPTAS if it results in an approximation ratio of $(1 + \varepsilon)$ and its running time is bounded by a polynomial in both the input size n and $1/\varepsilon$. The existence of FPTAS for a particular problem allows effective problem solving.
- **Polynomial-time approximation scheme (PAS or PTAS):** An approximation algorithm is a PTAS if it results in an approximation ratio of $(1 + \varepsilon)$ and its running time is bounded by a polynomial in the problem size n. Because the running time of an PTAS can be bounded by a polynomial with the exponent $1/\varepsilon$, an approximation scheme whose running time is $O(n^{1/\varepsilon})$ is still a PTAS. Thus, the running time of PTAS can increase polynomially with the input size n but exponentially with $1/\varepsilon$.
- **Constant-factor approximations (often abbreviated as APX for approximable):** Such approximation algorithms guarantee a constant approximation ratio and their running time is bounded by a polynomial in the problem size n.

FPTAS are the most effective approximation algorithms followed by PTAS and APX. FPTAS and PTAS are approximation schemes and we have a trade-off between computational effort and solution quality. Therefore, we are able to find better solutions if we allow a higher computational effort.

We can use the different types of approximation algorithms to extend our classification from Sect. 2.4.1 which distinguishes between easy problems (problems that can be solved in polynomial time) and NP-hard problems (problems where only exponential-time algorithms exist). We introduce some new complexity classes that are "between" P and NP and establish a relation between different types of problems:

$$P \subseteq FPTAS \subseteq PTAS \subseteq APX \subseteq NP$$

We assume that $P \neq NP$. In analogy to the definition of NP-hard problems, we can define APX-hard, PTAS-hard, and FPTAS-hard problems. If we are able to prove that no algorithm with polynomial running time exists for a particular problem, then the problem is at least FPTAS-hard. Analogously, problems are PTAS-hard if no FPTAS exists or APX-hard if it can be proven that no PTAS exists.

Figure 3.20 gives an overview of the different classes and lists some representative combinatorial optimization problems. We know for the general variant of the TSP problem that no approximation methods with polynomial running time can be developed (Orponen and Mannila, 1987). If the distances between the cities are symmetric, then the TSP becomes APX-complete (Papadimitriou and Yannakakis, 1993) and constant-factor approximations are possible. A representative example of a constant-factor approximation for symmetric TSPs is the Christofides heuristic which is discussed in the next paragraphs. Other examples of APX-hard problems are the *maximum satisfiability problem* (MAX SAT, see Sect. 4.4, p. 126), for which a 3/4-approximation ($\rho(n) = 4/3$) exists (Yannakakis, 1994) and *vertex cover*, which has a 2-approximation ($\rho(n) = 2$) (Papadimitriou and Steiglitz, 1982). The goal of MAX SAT is to assign variables of a given Boolean formula in such a way that the formula becomes true. Vertex cover aims at finding, for a given graph, the smallest set of vertices that is incident to every edge in the graph.

If the distances between the cities of a TSP are Euclidean, then PTASs exist (Arora, 1998). For the two-dimensional case where the cities are located on a two-dimensional grid, Arora developed a PTAS that finds a $(1 + \varepsilon)$-approximation to the optimal tour in time $O(n(\log n)^{O(1/\varepsilon)})$. For the d-dimensional problem, the running time increases to $O(n(\log n)^{(O(\sqrt{d}/\varepsilon))^{d-1}})$, which is nearly linear for any fixed ε and d (Arora et al, 1998).

Finally, the *knapsack problem* is a representative example for FPTAS-hard problems. In a knapsack problem, we have a set of n objects, each with a size and a profit. Furthermore, we have a knapsack of some capacity. The goal is to find a subset of objects whose total size is no greater than the capacity of the knapsack and whose profit is maximized. This problem can be solved using dynamic programming and there are fast FPTAS that can obtain a $1 + \varepsilon$ solution in time $O(n \log(1/\varepsilon) + 1/\varepsilon^4)$ (Lawler, 1979; Ibarra and Kim, 1975).

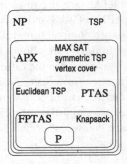

Fig. 3.20 Different classes of
NP optimization problems

In Sect. 3.4.1, we presented heuristics for the TSP and studied bounds on the expected solution quality. The heuristics presented are examples of approximation algorithms if their worst-case running time is polynomial and we are able to calculate a bound on their approximation ratio. Another prominent example of approximation algorithms for symmetric TSP is the *Christofides heuristic* (Christofides, 1976) which is also known as Christofides algorithm.

The Christofides heuristic was one of the first approximation algorithms illustrating that it is possible to develop effective heuristics for intractable problems. It accelerated a paradigm shift in the computer science and OR communities from trying to find exact methods to effective approximation algorithms. The heuristic is a constant-factor approximation with an approximation ratio of $\rho(n) = 3/2$. Its running time is $O(n^3)$.

The Christofides heuristic solves the TSP by combining a minimum spanning tree problem with the minimum weight perfect matching problem. Let $G = (V, E)$ be a connected, undirected graph with $n = |V|$ cities (nodes) and $m = |E|$ edges. The distance weights between every pair of nodes correspond to the city distances. The heuristic works as follows:

1. A *minimum weight tree* is a spanning tree that connects all the vertices and where the sum of the edge weights is minimal. Find a minimum weight tree $T = (V, R)$ (using for example Kruskal's algorithm). The *degree of a node* counts the number of edges that are connected to that node. Let $V_1 \subset V$ be those nodes having odd degree.
2. A *perfect matching* is a subset of edges without common vertices that touches all vertices exactly once. Any perfect matching of a graph with n nodes has $n/2$ edges. Find a minimum weight perfect matching M of V_1.
3. An *Euler tour* in an undirected graph is a cycle that uses each edge exactly once. Find an Euler tour S in the graph $(V, R \cup M)$.
4. Convert S into a TSP tour by taking shortcuts. That is, we obtain the tour by deleting from S all but one copy of each vertex in V.

Although Christofides's construction heuristic is more than 30 years old, it is still state-of-the-art and no better approximation algorithms for metric TSPs are known. Many researchers have conjectured that a heuristic with a lower approximation factor than $3/2$ may be achievable (Vazirani, 2003). However, there can be no polynomial-time algorithm with an approximation factor better than $220/219$ (Papadimitriou and Vempala, 2000).

For more details and information on approximation algorithms, we refer the reader to Hochbaum (1996) and Vazirani (2003) where an overview of different types of approximation algorithms for a variety of different, relevant optimization problems is given.

3.4.3 Modern Heuristics

Modern heuristics are extended variants of improvement heuristics (Reeves, 1993; Rayward-Smith et al, 1996; Rayward-Smith, 1998; Michalewicz and Fogel, 2004). They are general-purpose solvers and usually applicable to a large variety of problems. Modern heuristics are also called *heuristic optimization methods* or *meta-*

heuristics. In contrast to standard improvement heuristics which usually only perform improvement steps, modern heuristics also allow inferior solutions to be generated during search. Therefore, we want to define modern heuristics as improvement heuristics that

1. can be applied to a wide range of different problems (they are general-purpose methods) and
2. use during search both intensification (exploitation) and diversification (exploration) phases.

The goal of *intensification* steps is to improve the quality of solutions. In contrast, *diversification* explores new areas of the search space, thus accepting also complete or partial solutions that are inferior to the currently obtained solution(s).

We have chosen this definition of modern heuristics as studying the literature revealed that the existing naming conventions and definitions are inaccurate and often inconsistent. Often, modern heuristics are either defined very generally and abstractly as '*general-purpose methods that are very flexible and can be applied to many types of problems*' or by just enumerating different methods that are denoted as modern heuristics such as evolutionary algorithms, simulated annealing, tabu search, ant colony optimization, adaptive agents, and others. Both types of definitions are inaccurate and fuzzy and it remains unclear what are the differences between heuristics and modern heuristics. Defining modern heuristics as general-purpose improvement heuristics that use intensification and diversification phases allows a more precise distinction.

It is common to many modern heuristics that they mimic successful strategies found in other domains like for example nature, societies, or physics. Using this perspective, a modern heuristic defines a concept for controlling a heuristic search that can be applied to different optimization problems with relatively few modifications. It is characteristic of modern heuristics that their functionality is problem-independent, making it easy to apply such methods to different application domains. The ease of use of modern heuristics has resulted in a huge amount of different applications and there are few problems to which modern heuristics have not yet been applied. For overviews of the application of modern heuristics, we refer to selected literature (Biethahn and Nissen, 1995; Osman and Laporte, 1996; Bäck et al, 1997; Alander, 2000; Blum and Roli, 2003).

As modern heuristics are based on improvement heuristics, their functionality is similar. Usually, we start with one or more randomly generated complete solutions of a problem. Then, in iterative steps, we modify the existing solution(s) generating one or more new solutions. New solutions are created by search operators which are often called *variation operators*. Furthermore, we regularly perform intensification and diversification phases.

- During intensification, we use the objective value of the existing solutions and focus the application of variation operators on high-quality solutions.
- During diversification, we usually do not consider the objective value of solutions but systematically modify the existing solutions such that new areas of the search space are explored.

Usually, modern heuristics perform a limited number of search steps and are stopped after either a certain quality level is reached or a number of search steps are performed. For more details on the functionality of different types of modern heuristics, we refer to Chap. 5 and to selected literature (Rechenberg, 1973b; Holland, 1975; Schwefel, 1981; van Laarhoven and Aarts, 1988; Goldberg, 1989c; Fogel, 1995; Osman and Kelly, 1996; Osman and Laporte, 1996; Bäck, 1996; Michalewicz, 1996; Mitchell, 1996; Bäck et al, 1997; Glover and Laguna, 1997; Aarts and Lenstra, 1997; Goldberg, 2002; Ribeiro and Hansen, 2001; Langdon and Poli, 2002; Glover and Kochenberger, 2003; Blum and Roli, 2003; Reeves and Rowe, 2003; Resende and de Sousa, 2003; Gendreau, 2003; Hoos and Stützle, 2004; Alba, 2005; Burke and Kendall, 2005; Dréo et al, 2005; Gendreau and Potvin, 2005; Ibaraki et al, 2005; De Jong, 2006; Doerner et al, 2007; Siarry and Michalewicz, 2008).

The application of modern heuristics is easy as usually only two requirements are necessary.

- Representation: We must be able to represent complete solutions to a problem such that variation operators can be applied to them.
- Pairwise fitness comparisons: The *fitness function* returns the *fitness* of a solution. The *fitness function* indicates the quality of a solution and is based on, but is not necessarily identical to the objective function. For the application of modern heuristics, we must be able to compare the quality of two solutions and indicate which of the two solution has a higher objective value.

To successfully apply modern heuristics, usually it is not necessary to determine the correct, absolute objective value of a solution but pairwise comparisons of solutions are sufficient. If we can define an appropriate representation and perform pairwise fitness comparisons, we are able to apply a modern heuristic to a problem.

Studying the literature reveals that there exists a large variety of different modern heuristics. However, often the differences between the different approaches are small and similar concepts are presented using different labels. In the literature, various categorizations of modern heuristics have been proposed to classify the existing approaches. According to Blum and Roli (2003) the most common are to classify modern heuristics according to the origin of the modern heuristic (nature-inspired versus non-nature-inspired), the number of solutions that are considered in parallel during search (population-based versus single-point search), the type of fitness function (dynamic versus static fitness function), the type of neighborhood definition (static versus dynamic neighborhood definition), and the consideration of previous search steps for new solutions (memory usage versus memory-less). Although these classifications allow us to categorize modern heuristics, the categorizations are not necessarily meaningful with respect to the underlying principles of modern heuristics. If we want to design modern heuristics in a systematic way, we have to understand what are the basic ingredients of modern heuristics and how we have to combine them to obtain effective algorithms. Partial aspects, for example concerning the history of methods or the number of solutions used in parallel, are of minor importance. In the remainder of this work, we want to categorize modern heuristics using the following characteristics (main design elements):

1. **Representation**: The representation describes how different solutions of a problem are encoded. The representation must allow us to represent the optimal solution and we must be able to apply variation operators to the solutions.
2. **Variation operator(s)**: Variation operators are closely related to the representation used. They are iteratively applied to represented solutions and generate new solutions of a problem. We can distinguish between search operators that generate solutions that are similar to an existing solution and search operators that use a set of existing solutions and create a new solution by recombining relevant properties of those solutions.
3. **Fitness function**: Often, the fitness of a solution is identical to its objective value. Extensions are necessary if for example the search space contains infeasible solutions. Then, we have to assign fitness values not only to feasible but also to infeasible solutions. The use of modified or simplified fitness functions is possible as long as it allows us to distinguish high-quality from low-quality solutions.
4. **Initial solution(s)**: In many variants of modern heuristics, the initial solutions are created randomly. It is also possible to use an improvement heuristic to construct either one or a set of high-quality solutions which can be used as initial solution(s) for a modern heuristic.
5. **Search strategy**: The search strategy controls the sequence of intensification and diversification steps. Therefore, it also implicitly defines the intensification and diversification mechanisms. In principle, we can distinguish between approaches that use an explicit sequence of steps and approaches where intensification and diversification run in parallel.

These five aspects are the relevant design elements of modern heuristics. Furthermore, we can use them for the classification of different variants of modern heuristics. In the following chapters, we will discuss these aspects in detail and illustrate how we can systematically design effective and efficient modern heuristics.

In Sect. 2.4, we discussed properties of problems and presented measures for the locality and decomposability of a problem. Locality is a property of an optimization problem that is exploited by local search mechanisms, whereas decomposability is exploited by recombination-based search. Consequently, we can distinguish between modern heuristics using variation operators that are either based on local search or on problem decomposition. Local search methods use variation operators that generate new solutions that are similar to existing solutions. In contrast, recombination-based search methods generate new solutions by decomposing solutions and recombining building blocks to form new solutions.

We present *simulated annealing* (SA) as a representative example of modern heuristics. SA was independently proposed by Kirkpatrick et al (1983) and Cerny (1985) and is a local search method that can be used for continuous as well as combinatorial optimization problems. The fundamental idea of SA is to accept new, inferior solutions during the run to escape from local optima. The probability of accepting worse solutions depends on the solution quality and is reduced during a run. SA has its origins in the work of Metropolis et al (1958) and is called "annealing" as it imitates the crystallization process of molten liquids or metals. In metallurgy, annealing denotes a technique that heats up and slowly cools down liquid metals.

Heating up a solid material allows atoms to move around. If the cooling process is sufficiently slow, the size of the resulting crystals is large and few defects occur in the crystalline structure. Thus, crystal structures with a minimum energy configuration adopted by a collection of atoms are found.

SA imitates this behavior and uses iterative search steps. In each search step, a variation operator is applied to a solution x^o. For combinatorial optimization problems, the variation operator is often designed in such a way that it creates a neighboring solution x^n. Usually, the fitness $f(x)$ of a solution x is equivalent to its objective value and the search starts with a randomly created initial solution. The search strategy consists of intensification and diversification elements as the probability of accepting better solutions changes during the search process. The probability is controlled by the strategy parameter T which is often called temperature. At the beginning of the search, diversification dominates and the SA explores the search space (see Fig. 3.21(a)). New solutions x^n replace the old solutions x^o with high probability and also solutions with lower fitness are accepted. As the number of search steps increases, exploration becomes weaker and at the end of an SA run ($T \approx 0$) only better solutions are allowed to replace the current solution. The level of intensification and exploration is controlled by the acceptance probability $P_{acc}(T)$ of replacing a current solution x^o by a neighboring solution x^n. P_{acc} depends on the temperature T which is reduced during search and on the fitness difference $\delta E = f(x^n) - f(x^o)$ between old and new solution. For minimization problems, it is calculated as

$$P_{acc}(T) = \begin{cases} 1 & \text{if} \quad f(x^n) \leq f(x^o) \\ \exp(\frac{-\delta E}{T}) & \text{if} \quad f(x^n) > f(x^o), \end{cases}$$

where T is reduced during an SA run according to a cooling schedule. Figure 3.21(b) plots how P_{acc} depends on δE for different T and Algorithm 1 outlines the basic functionality of SA.

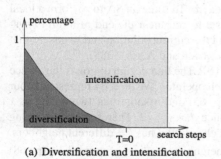

(a) Diversification and intensification

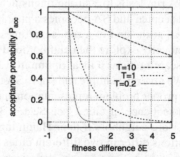

(b) Acceptance probability P_{acc} over fitness difference δE for different T

Fig. 3.21 Simulated annealing

Algorithm 1 Simulated Annealing (minimization problem)

Create initial solution x^o with objective value $f(x^o)$
Choose initial temperature T
repeat
   create neighboring solution x^n with objective value $f(x^n)$
   $\delta E = f(x^n) - f(x^o)$
   if $\delta E \leq 0$ **then**
     $x^o := x^n$
   else
     if $random[0,1) < \exp(-\frac{\delta E}{T})$ **then**
       $x^o := x^n$
     end if
   end if
   reduce T according to cooling schedule
until termination

The definition of a proper cooling schedule is one of the most important method-specific design options (van Laarhoven and Aarts, 1988; Aarts et al, 1997). If T is reduced very slowly, SA returns the optimal solution at the end of a run (Aarts and van Laarhoven, 1985; Aarts et al, 1997; Henderson et al, 2003). However, in practice, cooling schedules that guarantee finding the optimal solution are too slow and time-consuming as the necessary number of search steps often exceeds the size of the search space. Therefore, often a fixed cooling schedule is used, where the temperature at time $i+1$ is set to $T_{i+1} = cT_i$ $(0 < c < 1)$. Typical values are $c \in [0.9, 0.999]$. The initial temperature T_0 should be set such that exploration is possible at the beginning of the search and also some areas of the search space with low-quality solutions are explored. A proper strategy for setting T_0 is to randomly generate a number of solutions before the SA run and to set the initial temperature to $T_0 \approx \sigma(f(x)) \ldots 2\sigma(f(x))$, where $\sigma(f(x))$ denotes the standard deviation of the objective value of the randomly generated solutions. A low value of T at the end of an SA run leads to a strong intensification of the search. Then, only better solutions are accepted and SA becomes a pure local search. To force an SA to perform a local search at the end of a run, sometimes T is set to zero near the end of a run to systematically explore the neighborhood around the current solution and to ensure that the search returns a local optimum (van Laarhoven and Aarts, 1988).

We give an example and apply SA to the TSP. The first design option is the choice of a proper problem representation and search operator. We directly represent a tour as a sequence of cities (see Sect. 3.3.2.3, p. 65) and assume that two solutions x and y are neighbors if their Hamming distance $d(x,y) = 2$. Therefore, neighboring solutions differ in two different cities and each solution has $\binom{n}{2}$ different neighbors. If the initial city is fixed, the number of neighbors is $\binom{n-1}{2}$. Figure 3.14 (p. 68) illustrates a search space for a TSP with four cities where each solution has $\binom{3}{2} = 3$ neighbors (tour always starts with city a).

We have to determine a proper value for T_0. As the standard deviation of all fitness values is $\sigma = \sqrt{62}/3 \approx 2.62$ (we have three different solutions with fitness 11, 12, and 17), we set $T_0 = 3$. Furthermore, we use a simple, linear cooling schedule, where $T_{i+1} = 0.9T_i$. As in Fig. 3.14, we start with the initial solution $x^0 = abcda$ with $f(x^0) = 11$. The variation operator randomly exchanges the position of two cities in the tour and, for example, creates a random neighboring solution $x^1 = abdca$ with $f(x^1) = 17$. As $f(x^1) > f(x^0)$, x^1 replaces x^0 with probability $P = \exp(-6/3) \approx 0.14$. We generate a random number $rnd = [0, 1)$ and if $rnd < 0.14$, x^1 replaces x^0 and we continue with x^1. Otherwise, we continue with x^0. Then, we reduce the temperature T which becomes $T_1 = 2.7$. We continue iterations until we exceed a predefined number of search steps or until we have found no better solution for a certain number of search steps. With lowering T, the search converges with high probability to the optimal solution (there are only three solutions).

3.4.4 No-Free-Lunch Theorem

Already in its early days, the heuristic optimization community observed a general trade-off between effectiveness and application range of optimization methods (Polya, 1945; Simon and Newell, 1958; Kuehn and Hamburger, 1963; Romanycia and Pelletier, 1985). Often, the more problems could be solved with one particular optimization method, the lower its resulting average performance. Therefore, researchers have started to design heuristics in a more problem-specific way to increase the performance of heuristics for selected optimization problems.

The situation is different for modern heuristics as many of these methods are viewed as general-purpose problem solvers that reliably and effectively solve a large variety of different problems. The goal of many researchers in the modern heuristics field has been to develop modern heuristics that are black-box optimization methods. Black-box optimization methods are algorithms that need no additional information about the structure of a problem but are able to reliably and efficiently return high-quality solutions for a large variety of different optimization problems.

In 1995, Wolpert and Macready presented the No Free Lunch (NFL) theorem for optimization (Wolpert and Macready, 1995). It builds upon previous work (Watanabe, 1969; Mitchell, 1982; Radcliffe and Surry, 1995) and basically says that the design of general black-box optimization methods is not possible. In 1997, it was finally published in a journal (Wolpert and Macready, 1997). For an introduction, see Whitley and Watson (2005). The NFL theorem is concerned with the performance of algorithms that search through a search space by performing iterative search steps. The authors summarize the main result of the NFL theorem as:

> ... for both static and time dependent optimization problems, the average performance of any pair of algorithms across all possible problems is exactly identical. This means in particular that if some algorithm A_1's performance is superior to that of another algorithm A_2 over some set of optimization problems, then the reverse must be true over the set of all other optimization problems (Wolpert and Macready, 1997).

Therefore, black-box optimization is not possible. An algorithm's performance can only be high if (correct) problem-specific assumptions are made about the structure of the optimization problem and the algorithm is able to exploit these problem-specific properties.

We will have a closer look at the theorem. Wolpert and Macready assume a discrete search space X and an objective function f that assigns a fitness value $y = f(x)$ to each $x \in X$. During search, a set of m distinct solutions $\{d_m^x(i)\}$ with corresponding objective values $\{d_m^y(i)\}$ is created, where $i \in [1, \ldots, m]$ denotes an order on the sets (for example the order in which the solutions are created). New solutions are created step by step and after $m - 1$ search steps, we have the sets $\{d_m^x(1) = x^1, d_m^x(2) = x^2, \ldots, d_m^x(m) = x^m\}$ and $\{d_m^y(1) = f(x^1), d_m^y(2) = f(x^2), \ldots, d_m^y(m) = f(x^m)\}$. In each search step, a search algorithm A uses the previously generated solutions to decide which solution is created next. We assume that newly generated solutions have not been generated before. Therefore, A is a mapping from a previously visited set of solutions to a single new, unvisited solution, $A : \{d_m^x(i)\} \rightarrow x^{m+1}$, where $x^{m+1} \in X - \{d_m^x(i)\}$ and $i \in [1, \ldots, m]$. Usually, A considers the objective values of the previously visited solutions.

Wolpert and Macready use two performance measures. First, the performance of A can be measured using a function $\Phi(d_m^y)$ that depends on the set of objective values $d_m^y = \{d_m^y(i)\}$. $\Phi(d_m^y)$ can for example return the lowest fitness value in d_m^y: $\Phi(d_m^y) = \min_i\{d_m^y(i) : i = 1, \ldots, m\}$. Second, the performance of an algorithm A is measured using $P(d_m^y | f, m, A)$. This is the conditional probability of obtaining an ordered set of m distinct objective values $\{d_m^y(i)\}$ under the stated conditions.

The NFL theorem compares the performance of different algorithms A averaged over all possible f. Measuring an algorithm's performance using $P(d_m^y | f, m, A)$, the theorem says:

Theorem 3.1. *For any pair of algorithms A_1 and A_2,*

$$\sum_f P(d_m^y | f, m, A_1) = \sum_f P(d_m^y | f, m, A_2),$$

where the sum is over all possible functions f.

Therefore, the probability of generating a particular sequence of solutions is the same for any two search algorithms averaged over all possible objective functions. A direct corollary of this result is that for any performance measure $\Phi(d_m^y)$, the average of $P(\Phi(d_m^y) | f, m, A)$ over all f is also independent of A.

$$\sum_f P(\Phi(d_m^y) | f, m, A_1) = \sum_f P(\Phi(d_m^y) | f, m, A_2).$$

Therefore, there is no algorithm that outperforms some other algorithm on all problems that can be created by all possible assignments of objective values to solutions. If one algorithm gains in performance on one class of problems, it necessarily pays for it on the remaining problems.

Important for the interpretation of the NFL theorem is the definition of "*all possible (objective) functions f*". Schumacher (2000) and Schumacher et al (2001) introduced the *permutation closure* of a set of functions. We assume that we have an objective function $f : X \to Y$ where $f(x) = y$ assigns a $y \in Y$ to each $x \in X$ and X and Y are finite sets. $\sigma : X \to X$ is a permutation of all $x \in X$ and $\sigma^{-1} : X \to X$ is the inverse function, where $\sigma(\sigma^{-1}(x)) = x$. Consequently, we can permute functions as follows:

$$\sigma f(x) = f(\sigma^{-1}(x))$$

Therefore, a permutation $\sigma f(x)$ is equivalent to a permutation of all $x \in X$. Schumacher defined the *permutation closure* $P(F)$ of a set of functions F as:

$$P(F) = \{\sigma f : f \in F \text{ and } \sigma \text{ is a permutation}\}$$

$P(F)$ is constructed by taking a function f which assigns objective values $y = f(x)$ to all $x \in X$ and changing the assignment from $x \in X$ to $y \in Y$. Considering all possible functions f results in a closed set since every re-assignment of the mapping from $x \in X$ to $y \in Y$ will produce a function that is already a member of $P(F)$. Thus, $P(F)$ is *closed under permutation*. Schumacher et al (2001) proved the following result:

Theorem 3.2. *The NFL theorem holds for a set of functions if and only if that set of functions is closed under permutation.*

This means that the performance of different search heuristics is the same when averaged over a set of functions that is closed under permutation. The set of functions contains all objective functions that can be created when assigning the objective values in a different way to the solutions. Given a search space X with size $|X|$, Y with size $|Y| = |X|$, and an injective function f that assigns each $x \in X$ to a $y \in Y$, there are $|X|!$ different functions in the permutation closure. Figure 3.22 plots three out of $4! = 24$ different functions of the permutation closure for $|X| = |Y| = 4$. Averaged over all 24 possible objective functions, all search methods show the same performance.

Fig. 3.22 Three functions from a set of 24 functions closed under permutation

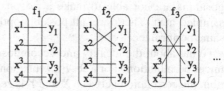

Important for a proper interpretation of the NFL theorem is which problems are closed under permutation. Trivial examples are problems where all solutions $x \in X$ have the same fitness value y. Then, obviously all search algorithms show the same performance. English (2000) recognized that NFL also holds for needle-in-a-haystack (NIH) problems ((2.12), p. 30). In such problems, all solutions have the same evaluation value except one which is the optimum and has a higher objective value (maximization problem). For the NIH, $|X| \neq |Y|$. Since $|Y| = 2$, there are only $|X|$ different functions in the permutation closure. The performance of search algorithms is the same when averaged over all $|X|$ functions. Figure 3.23 shows all possible NIH functions for $|X| = 4$. We assume a maximization problem and $y_1 > y_2$.

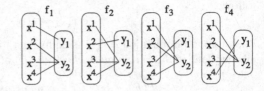

Fig. 3.23 The permutation closure of NIH problems with $|X| = 4$

NFL-like results can also be observed for problems that are not closed under permutation. Whitley and Rowe (2008) showed that a subset of algorithms can have identical performance over a subset of functions, even when the subset is not closed under permutation. In the extreme case, two algorithms can have identical performance over just two functions. In contrast to Theorem 3.1, which assumes unknown algorithms A_1 and A_2, Whitley and Rowe (2008) assume that there is some a-priori knowledge about the algorithms used. For example, they observe identical performance of algorithms on some functions that are not closed under permutation, if the search algorithms are limited to m steps, where m is significantly smaller than the size of the search space.

The consequences of the NFL theorem are in agreement with Sects. 2.3.2 and 2.4.2. In Sect. 2.3.2, we discussed trivial topologies where no meaningful neighborhood can be defined. Since all solutions are neighbors of each other, step-wise search algorithms are not able to make a meaningful guess at the solution that should be sampled next but just select a random solution from the search space. Therefore, the performance of all search algorithms is, on average, the same.

In Sect. 2.4.2.1, we discussed problems where no correlation exists between the distance between solutions and their corresponding objective values (see Fig. 2.3, middle). The situation is analogous to problems with a trivial topology. Search algorithms are not able to make a reasonable guess at the solution that should be sampled next. Therefore, all possible optimization methods show, on average, the same performance.

What about the locality and decomposability of a set of functions that is closed under permutation? For a set F of functions $f : X \to Y$ that is closed under permutation, the correlation between X and Y is zero averaged over all functions. Therefore, the locality, averaged over all functions, is low and the resulting functions can, on average, not be decomposed. We illustrate this for two types of problems. Measuring the correlation $\rho_{X,Y}$ between X and Y for NIH problems reveals that $\sum_F \rho_{X,Y}(f) = 0$. NIH problems have low locality and cannot be decomposed. The same holds for other problems that are closed under permutation like $f : X \to Y$ with $X = \{1, 2, 3\}$ and $Y = \{1, 2, 3\}$ (Fig. 3.24). The average correlation $\sum_F \rho_{X,Y}(f)$ between X and Y is $\sum_i \rho(f_i) = 1 + 0.5 + 0.5 - 0.5 - 1 - 0.5 = 0$.

We can generalize this result. As discussed in Sect. 2.2, a problem defines a set of problem instances. Each problem instance can be viewed as one particular function $f : X \to Y$. The performance of a search algorithm is the same averaged

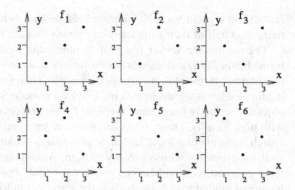

Fig. 3.24 Six functions closed under permutation

over all functions f that are defined by the problem if the functions are closed under permutation. Then, also the correlation $\rho_{X,Y}$ averaged over all functions f is zero.

Given a set of functions that are closed under permutation, search performance is also the same if we choose only a selection of functions from the permutation closure as long as the correlation $\rho_{X,Y}$ averaged over the selected functions is zero. Choosing a set of random functions from a permutation closure also results in the same performance of search algorithms on the selected set. Instead of using $\rho_{X,Y}$, we could also use the fitness-distance correlation ρ_{FDC} or other measurements that measure a correlation between X and Y.

After discussing types of problems for which we are not able to develop well-performing search algorithms, we might ask for which problems the NFL does not hold and high-performing heuristics become possible. Christensen and Oppacher (2001) studied the performance of an algorithm called *submedian-seeker*. The algorithm proceeds as follows (minimization problem):

1. Evaluate a sample of solutions and estimate $median(f)$ of the objective value.
2. If $f(x^i) < median(f)$, then sample a neighbor of x^i. Else sample a new random solution.
3. Repeat step 2 until half of the search space is explored.

We assume a metric search space where f is one-dimensional and a bijection. The number of submedian values of f that have supermedian successors is denoted as $M(f)$. For $M(f) = 1$ the set of solutions can be split into two halves, one with objective values lower than $median(f)$ and one with higher objective values. With increasing $M(f)$ the number of local optima increases. For $M(f) = |X|/2$, there are no neighboring solutions that both have an objective value below $median(f)$. Christensen and Oppacher (2001) showed that there exists M_{crit} such that when $M(f) < M_{crit}$ the proposed algorithm (submedian-seeker) beats random search. The work of Christensen and Oppacher was the origin of work dealing with algorithms that are *robust*: this means they are able to beat random search on a wide range of problems (Streeter, 2003; Whitley et al, 2004; Whitley and Rowe, 2006). Whitley et al (2004) generalized the submedian-seeker of Christensen and Oppacher and introduced algorithms with *subthreshold-seeking behavior*. Such algorithms spend more than half of their time sampling points that are below threshold.

The submedian-seeker (as well as other subthreshold-seeking algorithms) performs better than random search if solutions that have an objective value lower than the median (threshold) are grouped together. Therefore, local search methods (submedian seeker is a local search method as it generates neighboring solutions if their objective value is lower than $median(f)$) outperform random search for problems with high locality, where good solutions are grouped together (Sect. 2.4.2). Guided search methods like local search or submedian-seeker exploit locality and perform well for problems whose locality is high. Analogously, if problem locality is low, there is no correlation between solutions X and objective values Y and the performance of guided search methods is the same as random search.

Therefore, we are able to design search methods that outperform other methods like random search if the correlation $\rho_{X,Y}$ over a set of functions is unequal to zero. In the real world, two properties of problems are relevant: locality and decomposability (see Sects. 2.4.2 and 2.4.3). Both properties lead to optimization problems with problem instances f for which $\sum_f \rho_{X,Y} \neq 0$. Thus, the good performance of heuristic search methods that is observed for many problems in the real world can be explained by the fact that guided search methods exploit the high locality of a problem and recombination-based search methods exploit how well a problem can be decomposed into subproblems. As most of real-world problems are not closed under permutation and $\sum_f \rho_{X,Y} \neq 0$, heuristic optimization methods can beat other, non-problem-specific methods if they exploit locality, decomposability, or other problem properties.

Part II
Modern Heuristics

Chapter 4
Design Elements

In the previous chapters, we reviewed properties of optimization problems and representative examples of optimization methods for combinatorial optimization problems, including modern heuristics. We defined modern heuristics as improvement heuristics that can be applied to a wide range of different problems, and that use during search both intensification and diversification phases (see Sect. 3.4.3).

This chapter focuses on the main design elements of modern heuristics, namely the representation, search operator, fitness function, and initial solution. These design elements are relevant for all different types of modern heuristics and understanding the general concepts behind the design elements is a prerequisite for developing high-quality modern heuristics. The fifth design element of modern heuristics, which is the search strategy, is discussed in Chap. 5, along with a classification of modern heuristics. The search strategy defines the types of intensification and diversification mechanisms. In Chap. 6, we discuss how we can use knowledge about an optimization problem for the design of problem-specific modern heuristics. Problem-specific modern heuristics allow us to find high-quality solutions for larger problem instances with reasonable effort.

Representations are mappings that assign all problem solutions (phenotypes) to, usually linear, strings (genotypes). Given the genotypes, we can define a search space either by defining search operators or by selecting a metric (see Sect. 2.3.1). The definition of search operators determines the distances between genotypes, and vice versa. Solutions that are created by a local search operator are neighbors. Recombination operators generate offspring, where the distances between offspring and parents are usually equal to or smaller than the distance between parents. Therefore, the definition of search operators directly implies a neighborhood structure on the search space. Alternatively, we can first select a metric for the genotypes and then define the search operators. Given a metric, we must define local search operators such that they generate neighboring solutions. Analogously, recombination operators must generate offspring which are similar to both parents. Therefore, representation and operator depend on each other and cannot be decided independently of each other. For a more detailed analysis of representations that goes beyond the scope of this book we refer the interested reader to Rothlauf (2006).

F. Rothlauf, *Design of Modern Heuristics*, Natural Computing Series,
DOI 10.1007/978-3-540-72962-4_4, © Springer-Verlag Berlin Heidelberg 2011

Designing a fitness function and initialization method is usually easier than designing proper representations and search operators. The fitness function is determined by the objective function and allows modern heuristics to perform pairwise comparisons between solutions. Initial solutions are usually randomly created if no a priori knowledge about the problem exists.

This chapter starts with a guideline on how to find the right optimization method for a particular problem. Then, we discuss the different design elements. Section 4.2 discusses properties of representations and gives an overview of standard genotypes. Analogously, Sect. 4.3 presents recommendations for the design of search operators. Here, we distinguish between local and recombination search operators and present standard search operators with well-known behavior. Finally, Sects. 4.4 and 4.5 present guidelines for the design of fitness functions and initialization methods for modern heuristics.

4.1 Using Modern Heuristics

Since the literature provides a large number of different optimization methods, the identification of the "right" optimization method for a particular problem is usually a difficult task. We need some guidelines on how to find the right method.

Important for a proper choice of an optimization method is recognizing the fundamental and basic structure of a problem. For this process, we can use *problem catalogs* (Garey and Johnson, 1979; Vazirani, 2003; Crescenzi and Kann, 2003; Alander, 2000) and look for existing problems that are similar to the problem at hand. If the problem at hand is a standard problem and if it can be solved with polynomial effort by some optimization method (e.g. linear problems, Sect. 3.2), often such a method is the method of choice for our problem. In contrast, problem solving is more difficult if the problem at hand is NP-hard. In this case, problem solving becomes easier if either fast exact optimization methods or efficient approximation methods (e.g. fully polynomial-time approximation schemes, Sect. 3.4.2) are available. Modern heuristics are the method of choice if the problem at hand is NP-hard, difficult, and no efficient exact methods, approximation methods, or simple heuristics exist.

Relating a particular real-world problem to existing standard problems in the literature is sometimes difficult as problems in the real world usually differ from the well-defined problems that we can find in textbooks. Real-world problems often have additional constraints, additional decision variables, and other optimization goals. Therefore, the resulting problem models for "real" problems are large, complex, non-deterministic, and often different from standard problems from the literature. We can try to reduce complexity and make our problem more standard-like by removing constraints, simplifying the objective function, or limiting the number of decision variables (see Sect. 2.1.3). However, we often do not want to neglect some important aspects and are not happy with the simplified model. Furthermore, we never know if the reduced model is an appropriate model for the real world or if it is just an abstract model without any practical relevance. This problem was recognized early for example by Ackoff (1973, p. 670) who stated that '*accounts are given about how messes were murdered by reducing them to problems, how problems were murdered by reducing them to models, and how models were murdered by excessive exposure to the elements of mathematics.*' Therefore, when simplifying realistic models to make them similar to existing standard models we must be careful as the use of standardized models and methods easily leads to over-simplified models and "wrong" solutions for the original problem.

After formulating a realistic problem model, we have to solve it. Finding an appropriate modern heuristic for the problem at hand is a difficult task as we are confronted with two major difficulties: first of all, the literature is full of success stories of modern heuristics (for example Alander (2000)). When studying the literature, we get the impression that modern heuristics are fabulous optimization tools that can solve all possible types of problems in short time. Of course, experience as well as the NFL theorem tell us that such generalizations are wrong. However, the literature usually does not provide us with failures and limitations of modern heuristics but emphasizes successful applications. In many published applications, the optimization problems are chosen extremely carefully and often only limited comparisons to other methods are performed. Furthermore, the variables and design of the used modern heuristic is tweaked until its performance is sufficient to allow publication. When users try to apply the methods they find in the literature to their own real-world problems (which are usually slightly different and larger), they often observe low performance. Thus, they start changing and tweaking the parameters until the method delivers solutions of reasonable quality. This process is very time-consuming and frustrating for users that just want to apply modern heuristics to solve their problems.

The second major problem for users of modern heuristic is the choice and parameterization of modern heuristics. In the literature, we can find a huge variety of different modern heuristics we can choose from. Even more, there are many variants of the same search concepts which are denoted differently. Often, users have no overview of the different methods and have no idea which are the "right" methods for their problem. The situation becomes worse as in the literature mainly success stories are published (see previous paragraph) and limitations of approaches are often not known or remain fuzzy. Furthermore, existing comparisons in the literature of different types of modern heuristics are often biased due to the expertise and background of the authors. This is not surprising because researchers are usually able to find a better design and parameter setting for such types of modern heuristics with which they are more familiar. Therefore, as users are overwhelmed by the large variety of different methods, they most often choose out-of-the-box methods from standard text books. Applying such methods to complex and large real-world problems is often problematic and leads to low-quality results as users do not appropriately consider problem-specific knowledge for the design of the method. Considering problem-specific knowledge for the design of modern heuristics turns modern heuristics from black-box optimization methods into powerful optimization tools and is at the core of successful problem solving.

4.2 Representation

Successful and efficient use of modern heuristics depends on the choice of the genotypes and the representation - that is, the mapping from genotype to phenotype - and on the choice of search operators that are applied to the genotypes. These choices cannot be made independently of each other. The question whether a certain representation leads to a better performing modern heuristic than an alternative representation can only be answered when the operators applied are taken into account. The reverse is also true: deciding between alternative operators is only meaningful for a given representation.

In practice, one can distinguish two complementary approaches to the design of representations and search operators (Rothlauf, 2006). The first approach defines *representations* (also known as *decoders* or *indirect representations*) where a solution is encoded in a standard data structure, such as strings or vectors, and applies standard off-the-shelf search operators to these genotypes. To evaluate a solution, the genotype needs to be mapped to the phenotype space. The proper choice of this genotype-phenotype mapping is important for the performance of the search process. The second approach encodes solutions to the problem in its most "natural" problem space and designs search operators to operate on this search space. In this case, often no additional mapping between genotypes and phenotypes is necessary, but domain-specific search operators need to be defined. The resulting combination of representation and operator is often called *direct representation*.

This section focuses on representations. It introduces genotypes and phenotypes (Sect. 4.2.1) and discusses properties of the resulting genotype and phenotype space (Sect. 4.2.2). Section 4.2.3 lists the benefits of using (indirect) representations. Finally, Sect. 4.2.4 gives an overview of standard genotypes.

4.2.1 Genotypes and Phenotypes

In 1866, Mendel recognized that nature stores the complete genetic information for an individual in pairwise alleles (Mendel, 1866). The genetic information that determines the properties, appearance, and shape of an individual is stored by a number of strings. Later, it was discovered that the genetic information is formed by a double string of four nucleotides, called DNA.

Mendel realized that nature distinguishes between the genetic code of an individual and its outward appearance. The genotype represents all the information stored in the chromosomes and allows us to describe an individual on the level of genes. The phenotype describes the outward appearance of an individual. A transformation exists – a genotype-phenotype mapping or a representation – that uses the genotype information to construct the phenotype. To represent the large number of possible phenotypes with only four nucleotides, the genotype information is not stored in the alleles itself, but in the sequence of alleles. By interpreting the sequence of alleles, nature can encode a large number of different phenotypes using only a few different types of alleles.

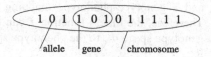

Fig. 4.1 Alleles, genes, and chromosomes

In Fig. 4.1, we illustrate the differences between *chromosome*, *gene*, and *allele*. A chromosome is a string of some length where all the genetic information of an individual is stored. Although nature often uses more than one chromosome, many modern heuristics use only one chromosome for encoding all phenotype information. Each chromosome consists of many alleles. Alleles are the smallest information units in a chromosome. In nature, alleles exist pairwise, whereas in most implementations of modern heuristics an allele is represented by only one symbol. For example, binary genotypes only have alleles with value zero or one. If a phenotypic property of an individual (solution), like its hair color or eye size is determined by one or more alleles, then these alleles together are called a gene. A gene is a region on a chromosome that must be interpreted together and which is responsible for a specific property of a phenotype.

We must carefully distinguish between genotypes and phenotypes. The phenotypic appearance of a solution determines its objective value. Therefore, when comparing the quality of different solutions, we must judge them on the phenotype level. However, when it comes to the application of variation operators we must view solutions on the genotype level. New solutions that are created using variation operators do not "inherit" the phenotypic properties of its parents, but only the genotype information regarding the phenotypic properties. Therefore, search operators work on the genotype level, whereas the evaluation of the solutions is performed on the phenotype level.

Formally, we define Φ_g as the genotype space where the variation operators are applied. An optimization problem on Φ_g could be formulated as

$$f(x) : \Phi_g \to \mathbb{R},$$

where f assigns an element (fitness value) in $\mathbb{R}$ to every element in the genotype space Φ_g. A maximization problem is defined as finding the optimal solution

$$x^* = \{x \in \Phi_g | \forall y \in \Phi_g : f(y) \leq f(x)\},$$

where x is usually a vector or string of decision variables (alleles) and $f(x)$ is the objective or fitness function. x^* is the global maximum. We have chosen a maximization problem, but without loss of generality, we could also model a minimization problem. To be able to apply modern heuristics to a problem, the inverse function f^{-1} does not need to exist.

4.2.2 Genotype and Phenotype Space

When using a representation, we have to define – in analogy to nature – genotypes and a genotype-phenotype mapping (Lewontin, 1974; Liepins and Vose, 1990). Therefore, the fitness function f can be decomposed into two parts. f_g maps the genotype space Φ_g to the phenotype space Φ_p, and f_p maps Φ_p to the fitness space $\mathbb{R}$:

$$f_g(x^g) : \Phi_g \to \Phi_p,$$
$$f_p(x^p) : \Phi_p \to \mathbb{R},$$

where $f = f_p \circ f_g = f_p(f_g(x^g))$. The genotype-phenotype mapping f_g is determined by the type of genotype used. f_p represents the fitness function and assigns a fitness value $f_p(x^p)$ to each solution $x^p \in \Phi_p$. The search operators are applied to the genotypes (Bagley, 1967; Vose, 1993).

We have seen in Sect. 2.3.1 that formulating a search space Φ also defines a metric. Using a metric, the distance $d(x, y)$ between two solutions $x, y \in \Phi$ measures how different the two solutions are. The larger the distance, the more different two individuals are with respect to the metric used. In principle, different metrics can be used for the same search space. Different metrics result in different distances and different measurements for the similarity of solutions.

Two individuals are neighbors if the distance between them is minimal. For example, when using the Hamming metric (2.7) for binary strings the minimal distance between two individuals is $d_{min} = 1$. Therefore, two individuals x and y are neighbors if their distance $d(x, y) = 1$.

Using a representation f_g, we obtain two different search spaces, Φ_g and Φ_p. Therefore, different metrics can be defined for the phenotype and the genotype space. The metric used on the phenotype search space Φ_p is usually determined by the specific problem to be solved and describes which problem solutions are similar to each other. Examples of common phenotypes and corresponding metrics are given in Sect. 4.3.5. In contrast, the metric defined on Φ_g is not defined by the specific problem but can be defined by us. We define a metric by defining either the genotypes or the corresponding search operators. As we can define different types of genotypes to represent the phenotypes, we are able to define different metrics on Φ_g. However, if the metrics on Φ_p and Φ_g are different, different neighborhoods can exist on Φ_g and Φ_p (Sect. 2.3.2). For example, when encoding phenotype integers using genotype bitstrings, the phenotype $x^p = 5$ has two neighbors, $y^p = 6$ and $z^p = 4$. When using Hamming metric and binary genotypes, the corresponding binary string $x^g = 101$ has three different neighbors, $y^g = 001$, $z^g = 111$, and $w^g = 100$ (Caruana and Schaffer, 1988).

Therefore, the metric on the genotype space should we chosen such that it fits well the metric on the phenotype space. A representation introduces an additional genotype-phenotype mapping and thus modifies the fit. We have to ensure that the metric on the genotype search space fits the original problem metric. We should choose the genotype metric in such a way that phenotypic neighbors remain neighbors in the genotype search space. Representations that ensure that neighboring phenotypes are also neighboring genotypes are called high-locality representations (see Sect. 6.1.2).

4.2.3 Benefits of Representations

In principle, a representation which defines an additional genotype-phenotype mapping is not necessary for the application of modern heuristics as search operators may also be directly applied to phenotypes. However, the use of representations has some benefits:

- The use of representations is necessary for problems where a phenotype cannot be depicted as a string or in another way that is accessible to variation operators. A representative example is the shape of an object, for example the wing of an airplane. Modern heuristics that are used to find the optimal shape of a wing usually require a representation as the direct application of search operators to the *shape* of a wing is difficult. Therefore, additional genotype-phenotype mappings are used and variation operators are applied to genotypes that indirectly determine the shape.
- The introduction of a representation can be useful if there are constraints or restrictions on the phenotype space that can be advantageously modeled by a specific encoding. An example is a tree problem where the optimal solution is a star. Instead of applying search operators directly to trees, we can introduce genotypes that only encode stars resulting in a much smaller search space.
- The use of the same genotypes for different types of problems, and only interpreting them differently by using a different genotype-phenotype mapping, allows us to use standard search operators (Sect. 4.3.5) with known properties. In this case, we do not need to develop any new operators with unknown properties and behavior.
- Finally, using an additional genotype-phenotype mapping can change the difficulty of a problem. The use of a representation can be helpful if such a mapping reduces problem difficulty and makes the problem easier to solve for a particular optimization method. However, usually the definition of a proper encoding is difficult and problem-specific.

4.2.4 Standard Genotypes

We describe some of the most important and widely used genotypes, and summarize some of their major characteristics. For a more detailed overview of different types of genotypes, we refer to Bäck et al (1997, Sect. C1).

4.2.4.1 Binary Genotypes

Binary genotypes are commonly used in genetic algorithms (Goldberg, 2002, 1989c). Such types of modern heuristics use recombination as the main search operator and mutation only serves as background noise. A typical search space is $\Phi_g = \{0,1\}^l$, where l is the length of a binary vector $x^g = (x_1^g, \ldots, x_l^g) \in \{0,1\}^l$. The genotype-phenotype mapping f_g depends on the specific optimization problem to be solved. For many combinatorial optimization problems using binary genotypes allows a direct and very natural encoding.

When using binary genotypes for encoding integer phenotypes, specific genotype-phenotype mappings are necessary. Different types of binary representations for integers assign the integers $x^p \in \Phi_p$ (phenotypes) in different ways to the binary vectors $x^g \in \Phi_g$ (genotypes). The most common binary genotype-phenotype mappings are binary, Gray, and unary encoding. For a more detailed description of these three types of encodings, we refer to Rothlauf (2006, Chap. 5) and Rowe et al (2004).

When using binary genotypes to encode continuous phenotypes, the accuracy (precision) depends on the number of bits that represent one phenotype variable. By increasing the number of bits that are used to represent one continuous variable the accuracy of the representation can be increased.

4.2.4.2 Integer Genotypes

Instead of using binary strings with cardinality $\chi = 2$ higher χ-ary alphabets, where $\chi \in \{\mathbb{N}^+ \setminus \{0,1\}\}$, can also be used for the genotypes. Then, instead of a binary alphabet a χ-ary alphabet is used for a string of length l. Instead of encoding 2^l different individuals with a binary alphabet, we are able to encode χ^l different possibilities. The size of the search space increases from $|\Phi_g| = 2^l$ to $|\Phi_g| = \chi^l$.

For many integer problems, users often prefer to use binary instead of integer genotypes because schema processing (Sect. 2.4.3.3) is maximally efficient with binary alphabets when using standard recombination operators in genetic algorithms (Goldberg, 1990). Goldberg (1991b) qualified this recommendation and emphasized that the alphabet used in the encoding should be as small as possible while still allowing a natural representation of solutions. To give general recommendations is difficult, as users often do not know a priori whether binary genotypes allow a natural encoding of integer phenotypes (Radcliffe, 1997; Fogel and Stayton, 1994). We recommend that users use binary genotypes for encoding binary decision variables and integer genotypes for integer decision variables.

4.2.4.3 Continuous Genotypes

When using continuous genotypes, the search space is $\Phi_g = \mathbb{R}^l$, where l is the size of a real-valued string or vector. Continuous genotypes are often used in local search methods like evolution strategies (Sect. 5.1.5) or evolutionary programming. These types of optimization are mainly based on local search and search through the search space by adding a multivariate zero-mean Gaussian random variable to each continuous variable. In contrast, when using recombination-based genetic algorithms continuous decision variables are often represented by using binary genotypes.

Continuous genotypes cannot only be used for encoding continuous problems, but also for permutation and combinatorial problems. Trees, schedules, tours, or other combinatorial problems can easily be represented by using continuous genotypes and special genotype-phenotype mappings (for examples see Sects. 8.1.2 and 8.4.1).

4.2.4.4 Messy Representations

In all previously presented genotypes, the position of each allele is fixed along the chromosome and only the corresponding value is specified. The first gene-independent genotype was proposed by Holland (1975). He proposed the inversion operator which changes the relative order of the alleles in the string. The position of an allele and the corresponding value are coded together as a tuple in a string. This concept can be used for all types of genotypes such as binary, integer, and real-valued alleles and allows an encoding which is independent of the position of the alleles in the chromosome. Later, Goldberg et al (1989) used this position-independent representation for the messy genetic algorithm.

4.3 Search Operator

This section discusses search operators. We want to distinguish between standard search operators which are applied to genotypes and problem-specific search operators that can also be applied to phenotypes (often called *direct representations*).

Section 4.3.1 starts with an overview of general design guidelines for search operators. Sections 4.3.2 and 4.3.3 discuss local search operators and recombination operators, respectively. In Sect. 4.3.4, we focus on direct representations. In direct representations, search operators are directly applied to phenotypes and no explicit genotype-phenotype mapping exists. Finally, Sect. 4.3.5 gives an overview of standard search operators for a variety of different genotypes.

4.3.1 General Design Guidelines

During the 1990s, Radcliffe developed guidelines for the design of search operators. It is important for search operators that the representation used is taken into account as search operators are based on the metric that is defined on the genotype space. Radcliffe introduced the principle of *formae*, which are subsets of the search space (Radcliffe, 1991b,a, 1992, 1993; Radcliffe and Surry, 1994; Radcliffe, 1994). Formae are defined as *equivalence classes* that are induced by a set of equivalence relations. Any possible solution of an optimization problem can be identified by specifying the equivalence class to which it belongs for each of the equivalence relations. For example, if we have a search space of faces (Surry and Radcliffe, 1996), basic equivalence relations might be "same hair color" or "same eye color", which would induce the formae "red hair", "dark hair", "green eyes", etc. Formae of higher order like "red hair and green eyes" are then constructed by composing simple formae. The search space, which includes all possible faces, can be constructed with strings of alleles that represent the different formae. For the definition of formae, the structure of the phenotypes is relevant. For example, for binary problems, possible formae would be "bit i is equal to one/zero". When encoding tree structures, possible basic formae would be "contains link from node i to node j".

It is an unsolved problem to find appropriate equivalences for a particular problem. From the equivalences, the genotype search space Φ_g and the genotype-phenotype mapping f_g can be constructed. Usually, a solution is encoded as a string of alleles. The value of an allele indicates whether the solution satisfies a particular equivalence. Radcliffe (1991a) proposed several design guidelines for creating appropriate equivalences for a given problem. The most important design guideline is that the generated formae should group together solutions of related fitness (Radcliffe and Surry, 1994), in order to create a fitness landscape or structure of the search space that can be exploited by search operators.

Radcliffe recognized that the genotype search space, the genotype-phenotype mapping, and the search operators belong together and their design cannot be separated from each other (Radcliffe, 1992). He assumed that search operators create offspring solutions from a set of parent solutions. For the development of appropriate search operators that are based on predefined formae, he formulated the following four design principles (Radcliffe, 1991a, 1994):

- **Respect**: Offspring produced by recombination should be members of all formae to which both their parents belong. This means for the "face example" that offspring should have red hair and green eyes if both parents have red hair and green eyes.
- **Transmission**: An offspring should be equivalent to at least one of its parents under each of the basic equivalence relations. This means that every gene should be set to an allele which is taken from one of the parents. If one parent has dark hair and the other red hair, then the offspring has either dark or red hair.
- **Assortment**: An offspring can be formed with any compatible characteristics taken from the parents. Assortment is necessary as some combinations of equivalence relations may be infeasible. This means for example, that the offspring inherits dark hair from the first parent and blue eyes from the second parent only if dark hair and blue eyes are compatible. Otherwise, the alleles are set to feasible values taken from a random parent.
- **Ergodicity**: An iterative use of search operators allows us to reach any point in the search space from all possible starting solutions.

The recommendations from Radcliffe confirm that representations and search operators depend on each other and cannot be designed independently. He developed a consistent concept of how to design efficient modern heuristics once appropriate equivalence classes (formae) are defined. However, the finding of appropriate equivalence classes, which is equivalent to either defining the genotype search space and the genotype-phenotype mapping or appropriate direct search operators on the phenotypes, is often difficult and remains an unsolved problem.

As long as the genotypes are either binary, integer, or real-valued strings, standard recombination and mutation operators can be used. The situation is different if direct representations (Sect. 4.3.4) are used for problems whose phenotypes are not binary, integer, or real-valued. Then, standard recombination and mutation operators cannot be used any more. Specialized operators are necessary that allow offspring to inherit important properties from their parents (Radcliffe, 1991a,b; Kargupta et al, 1992; Radcliffe, 1993). In general, these operators are problem-specific and must be developed separately for every optimization problem.

4.3.2 Local Search Operators

Local search and the use of local search operators are at the core of modern heuristics. The goal of local search is to generate new solutions with similar properties in comparison to the original solutions (Doran and Michie, 1966). Usually, a local search operator creates offspring that have a small or sometimes even minimal distance to their parents. Therefore, local search operators and the metric on the corresponding search space cannot be decided independently of each other but determine each other. A metric defines possible local search operators and a local search operator determines the metric. As search operators are applied to the genotypes, the metric on Φ_g is relevant for the definition of local search operators.

The basic idea behind using local search operators is that the structure of a fitness landscape should guide a search heuristic to high-quality solutions (Manderick et al, 1991), and that good solutions can be found by performing small iterated changes. We assume that high-quality solutions are not isolated in the search space but grouped together (Christensen and Oppacher, 2001; Whitley, 2002). Therefore, better solutions can be found by searching in the neighborhood of already found good solutions (see also the discussion on the submedian seeker in Sect. 3.4.4, p. 101). The search steps must be small because too large search steps would result in randomization of the search, and guided search around good solutions would become impossible. In contrast, when using search operators that perform large steps in the search space it would not be possible to find better solutions by searching around already found good solutions but the search algorithm would jump randomly around the search space (see Sect. 6.1).

The following paragraphs review some common local search operators for binary, integer, and continuous genotypes and illustrate how they are designed based on the underlying metric. The local search operators (and underlying metrics) are commonly used and usually a good choice. However, in principle we are free to choose other metrics and to define corresponding search operators. Then, the metric should be chosen such that high-quality solutions are neighboring solutions and the resulting fitness landscape leads guided search methods to an optimal solution. The choice of a proper metric and corresponding search operators are always problem-specific and the ultimate goal is to choose a metric such that the problem becomes easy for guided search methods (see the discussion on how locality affects guided search methods, Sect. 2.4.2). However, we want to emphasize that for most practical applications the illustrated search operators are a good choice and allow us to design efficient and effective modern heuristics.

4.3.2.1 Binary Genotypes

When using binary genotypes, the distance between two solutions $x, y \in \{0, 1\}^l$ is often measured using the Hamming distance (2.7). Many local search operators based on this metric generate new solutions with Hamming distance $d(x, y) = 1$. This type of search operator is also known as a *standard mutation operator* for binary strings or a *bit-flipping* operator. As each binary solution of length l has l neighbors, this search operator can create l different offspring. For example, applying the bit-flipping operator to $(0, 0, 0, 0)$ can result in four different offspring $(1, 0, 0, 0)$, $(0, 1, 0, 0)$, $(0, 0, 1, 0)$, and $(0, 0, 0, 1)$.

Reeves (1999b) proposed another local search operator for binary strings based on a different neighborhood definition: for a randomly chosen $k \in \{0, \ldots, l\}$, it complements the bits $x_k, \ldots, x_l$. Again, each solution has l neighbors. For example, applying this search operator to $(0, 0, 0, 0)$ can result in $(1, 1, 1, 1)$, $(0, 1, 1, 1)$, $(0, 0, 1, 1)$, or $(0, 0, 0, 1)$. Although the operator is of minor practical importance, it has some interesting theoretical properties. First, it is closely related to the one-point recombination crossover (see below) as it chooses a random point and inverts all x_i with $i \geq k$. Therefore, it has also been called the *complementary crossover operator*. Second, if all genotypes are encoded using Gray code (Gray, 1953; Caruana et al, 1989), the neighbors of a solution in the Gray-coded search space using Hamming distance are identical to the neighbors in the original binary-coded search space using the complementary crossover operator. Therefore, Hamming distances between Gray encoded solutions are equivalent to the distances between the original binary encoded solutions using the metric induced by the complementary crossover operator (neighboring solutions have distance one). For more information regarding the equivalence of different neighborhood definitions and search operators we refer to the literature (Reeves, 1999b; Höhn and Reeves, 1996a,b).

4.3.2.2 Integer Genotypes

For integer genotypes different metrics are common, leading to different local search operators. When using the binary Hamming metric (2.8), two individuals are neighbors if they differ in one decision variable. Search operators based on this metric assign a random value to a randomly chosen allele. Therefore, each solution $x \in \{0,\ldots,k\}^l$ has lk neighbors. For example, $x = (0,0)$ with $x_i \in \{0,1,2\}$ has four different neighbors $((1,0), (2,0), (0,1),$ and $(0,2))$.

The situation changes when defining local search operators based on the city-block metric (2.5). Then, a local search operator can create new solutions by slightly increasing or decreasing one randomly chosen decision variable. For example, new solutions are generated by adding +/-1 to a randomly chosen variable x_i. Each solution of length l has a maximum of $2l$ different neighbors. For example, $x = (1,1)$ with $x_i \in \{0,1,2,3\}$ has four different neighbors $((0,1), (2,1), (1,0),$ and $(1,2))$.

Finally, we can define search operators such that they do not modify values of decision variables but exchange values of two decision variables x_i and x_j. Therefore, using binary Hamming distance (2.8), two neighbors have distance $d = 2$ and each solution has a maximum of $\binom{l}{2}$ different neighbors. For example, $x = (3,5,2)$ has three different neighbors $((5,3,2), (2,5,3),$ and $(3,2,5))$.

4.3.2.3 Continuous Genotypes

For continuous genotypes, we can define local search operators analogously to integer genotypes. Based on the binary Hamming metric (2.8), the application of a local search operator can assign a random value $x_i \in [x_{i,min}, x_{i,max}]$ to the ith decision variable. Furthermore, we can define a local search operator such that it exchanges the values of two decision variables x_i and x_j. The binary Hamming distance between old and new solutions is $d = 2$.

The situation is a little more complex in comparison to integer genotypes when designing a local search operator based on the city-block metric (2.5). We must define a search operator such that its iterative application allows us to reach all solutions in reasonable time. Therefore, a search step should be not too small (we want to have some progress in search) and not too large (the offspring should be similar to the parent solution). A commonly used concept for such search operators is to add a random variable with zero mean to the decision variables. This results in $x'_i = x_i + m$, where m is a random variable and x' is the offspring generated from x. Sometimes m is uniformly distributed in $[-a,a]$, where $a < (x_{i,max} - x_{i,min})$. More common is the use of a normal distribution $\mathcal{N}(0,\sigma)$ with zero mean and standard deviation σ. The addition of zero-mean Gaussian random variables generates offspring that have, on average, the same statistical properties as their parents. For more information on local search operators for continuous variables, we refer to Fogel (1997).

4.3.3 Recombination Operators

To be able to use recombination operators, a set of solutions (which is usually called a *population*) must exist as the goal of recombination is to recombine meaningful properties of parent solutions. Therefore, for the application of recombination operators at least two parent solutions are necessary; otherwise local search operators are the only option. The design of recombination operators should follow the concepts proposed by Radcliffe (Sect. 4.3.1).

Analogously to local search operators, recombination operators should be designed based on the used metric (Liepins and Vose, 1990; Surry and Radcliffe, 1996). Given two parent solutions x^{p1} and x^{p2} and one offspring solution x^o, recombination operators should be designed such that

$$d(x^{p1}, x^{p2}) \geq \max(d(x^{p1}, x^o), d(x^{p2}, x^o)). \tag{4.1}$$

Therefore, the application of recombination operators should result in offspring where the distances between offspring and its parents are equal to or smaller than the distance between the parents. When viewing the distance between two solutions as a measurement of dissimilarity, this design principle ensures that offspring solutions are similar to parents. Consequently, applying a recombination operator to the same parent solutions $x^{p1} = x^{p2}$ should also result in the same offspring ($x^o = x^{p1} = x^{p2}$).

In the last few years, this basic concept of the design of recombination operators has been interpreted as "geometric crossover" (Moraglio and Poli, 2004; Moraglio et al, 2007; Moraglio, 2007). This work builds upon previous work (Liepins and Vose, 1990; Surry and Radcliffe, 1996; Rothlauf, 2002) and defines crossover and mutation representation-independently using the notion of distance associated with the search space.

Why should we use recombination operators in modern heuristics? The motivation is that we assume that many problems are decomposable (Sect. 2.4.3). Therefore, problems can be solved by decomposing them into smaller subproblems, solving these smaller subproblems, and combining the optimal solutions of the subproblems to obtain overall problem solutions. The purpose of recombination operators is to form new overall solutions by recombining solutions of smaller subproblems that exist in different parent solutions. If this juxtaposition of smaller, highly fit, partial solutions (often denoted as building blocks; Sect. 2.4.3.3) does not result in good solutions, search strategies that are based on recombination operators will lead to low performance. However, as many problems of practical relevance can be decomposed into smaller problems (they are decomposable), the use of recombination operators often results in good performance of modern heuristics.

The most common recombination operators for standard genotypes are *one-point crossover* (Holland, 1975) and *uniform crossover* (Reed et al, 1967; Ackley, 1987; Syswerda, 1989). We assume a vector or string x of decision variables of length l. When using one-point crossover, a *crossover point* $c = \{1, \ldots, l-1\}$ is initially chosen randomly. Usually, two offspring solutions are created from two parent solutions by swapping the partial strings. As a result, we get for the parents $x^{p1} = [x_1^{p1}, x_2^{p1}, \ldots, x_l^{p1}]$ and $x^{p2} = [x_1^{p2}, x_2^{p2}, \ldots, x_l^{p2}]$ the offspring $x^{o1} = [x_1^{p1}, x_2^{p1}, \ldots, x_c^{p1}, x_{c+1}^{p2}, \ldots, x_l^{p2}]$ and $x^{o2} = [x_1^{p2}, x_2^{p2}, \ldots, x_c^{p2}, x_{c+1}^{p1}, \ldots, x_l^{p1}]$. A generalized version of one-point crossover is *n-point crossover*. For this type of crossover operator, we choose n different crossover points and create an offspring by alternately selecting alleles from parent solutions. For uniform crossover, we decide independently for every single allele of the offspring from which parent solution it inherits the value of the allele. In most implementations, no parent is preferred and the probability of an offspring inheriting the value of an allele from a specific parent is $p = 1/m$, where m denotes the number of parents that are considered for recombination. For example, when two possible offspring are considered with the same probability ($p = 1/2$), we could get as offspring $x^{o1} = [x_1^{p1}, x_2^{p1}, x_3^{p2}, \ldots, x_{l-1}^{p1}, x_l^{p2}]$ and $x^{o2} = [x_1^{p2}, x_2^{p2}, x_3^{p1}, \ldots, x_{l-1}^{p2}, x_l^{p1}]$. We see that uniform crossover is equivalent to $(l-1)$-point crossover.

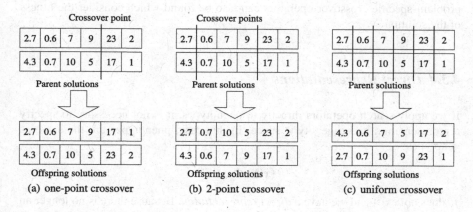

Fig. 4.2 Different crossover variants

Figure 4.2 presents examples for the three crossover variants. All three recombination operators are based on the binary Hamming distance and follow (4.1) as $d(x^{p1}, x^{p2}) \geq \max(d(x^{p1}, x^o), d(x^{p2}, x^o))$. Therefore, the similarity between offspring and parent is higher than between the parents.

Uniform and n-point crossover can be used independently of the type of decision variables (binary, integer, continuous, etc) since these operators only exchange alleles between parents. In contrast, *intermediate recombination* operators attempt to average or blend components across multiple parents and are designed for continuous and integer problems. Given two parents x^{p1} and x^{p2}, a crossover operator known as *arithmetic crossover* (Michalewicz, 1996) creates an offspring x^o as:

$$x_i^o = \alpha x_i^{p1} + (1 - \alpha) x_i^{p2},$$

where $\alpha \in [0,1]$. If $\alpha = 0.5$, the crossover just takes the average of both parent solutions. In general, for m parents, this operator becomes

$$x_i^o = \sum_{i=1}^{m} \alpha_i x_i^{pi},$$

where $\sum_{i=1}^{m} \alpha_i = 1$. Arithmetic crossover is based on the city-block metric (2.5). With respect to this metric, the distance between offspring and parent is smaller than the distance between the parents. Another type of crossover operator that is based on the Euclidean metric (2.6) is *geometrical crossover* (Michalewicz, 1996). Given two parents, an offspring is created as:

$$x_i^o = \sqrt{x_i^{p1} x_i^{p2}}.$$

For further information on crossover operators we refer to Booker (1997) (binary crossover) and Fogel (1997) (continuous crossover). There, a brief overview of problem-specific crossover operators can also be found which consider the fitness of the solution.

4.3.4 Direct Representations

If we apply search operators directly to phenotypes, it is not necessary to specify a representation and a genotype space. In this case, phenotypes are the same as genotypes:

$$f(x^g) : \Phi_g \rightarrow \mathbb{R}.$$

f_g does not exist and we have a *direct representation*. Because there is no longer an additional mapping between Φ_g and Φ_p, a direct representation does not change any aspect of the phenotype problem such as difficulty or metric. However, when using direct representations, we often cannot use standard search operators, but have to define problem-specific operators. Therefore, important for the success of modern heuristics using a direct representation is not finding a "good" representation for a specific problem, but developing proper search operators defined on phenotypes.

Relevant for different implementations of direct representations are not the representations used (there are no genotype-phenotype mappings) but the definition of the variation operators. Since we assume that local search operators always generate neighboring solutions, the definition of a local search operator induces a metric on the genotypes. Therefore, the metric that we use on the genotype space should be chosen in such a way that new solutions that are generated by local search operators have small (or better minimal) distance to the old solutions and the solutions are neighbors with respect to the metric used. Furthermore, the distance between two solutions $x \in \Phi_g$ and $y \in \Phi_g$ should be proportional to the minimal number of local search steps that are necessary to move from x to y. Analogously, the definition of a recombination operator also induces a metric on the search space. The metric used should guarantee that the application of a recombination operator to two solutions $x^p \in \Phi_g$ and $y^p \in \Phi_g$ creates a new solution $x^o \in \Phi_g$ whose distances to the parents are not larger than the distance between the parents (see equation (4.1)).

For the definition of variation operators, we should also consider that for many problems we have a natural notion of similarity between phenotypes. When we create a problem model, we often "know" whether two solutions are similar to each other, or not. Such a notion of similarity should be considered for the definition of variation operators. We should design local search operators in such a way that their application creates solutions which we "view" as similar. Such a definition of local search operators ensures that neighboring phenotypes are also neighbors with respect to the metric that is induced by the search operators.

At a first glance, it seems that the use of direct representations makes life easier as direct representations release us from the challenge to design efficient representations. However, we are confronted with some problems:

- To many phenotypes no standard variation operators can be applied.
- The design of high-quality problem-specific search operators is difficult.
- We cannot use use modern heuristics that only work on standard genotypes.

For indirect representations with standard genotypes, the definition of search operators is straightforward as these are usually based on the metric of the genotype space (see Sects. 4.3.2 and 4.3.3). The behavior of modern heuristics using standard search operators is usually well examined and well understood. However, when using direct representations, standard operators often can no longer be used. For each different phenotype, problem-specific operators must be developed. This is difficult, as we cannot use most of our knowledge about the behavior of modern heuristics using standard genotypes and standard operators.

The design of proper search operators is often demanding as phenotypes are usually not string-like but are more complicated structures like trees, schedules, or other structures (for some examples, see Sect. 4.3.5). In this case, phenotypes cannot be depicted as a string or in another way that is accessible to variation operators. Other representative examples are the form or shape of an object. Search operators that can be directly applied to the shape of an object are often difficult to design.

Finally, using specific variants of modern heuristics like estimation of distribution algorithms (EDA, Sect. 5.2.2) becomes very difficult. These types of modern heuristics do not use standard search operators that are applied to genotypes but build new solutions according to a probabilistic model of previously generated solutions (Mühlenbein and Paaß, 1996; Mühlenbein and Mahnig, 1999; Pelikan et al, 1999a,b; Larrañaga et al, 1999; Bosman, 2003). These search methods were developed for a few standard genotypes (usually binary and floats) and result in better performance than, for example, traditional simple genetic algorithms for decomposable problems (Larrañaga and Lozano, 2001; Pelikan, 2006). However, because direct representations with non-standard phenotypes and problem-specific search operators can hardly be implemented in EDAs, direct representations cannot benefit from these optimization methods.

4.3.5 Standard Search Operators

The following paragraphs provide an overview of standard search spaces and the corresponding search operators. The search spaces can either represent genotypes (indirect representation) or phenotypes (direct representation). We order the search spaces by increasing complexity. With increasing complexity of the search space, the design of search operators becomes more demanding. An alternative to designing complex search operators for complex search spaces is to introduce additional mappings that map complex search spaces to simpler ones. Then, the design of the corresponding search operators becomes easier, however, a proper design of the additional mapping (representation) becomes more important.

4.3.5.1 Strings and Vectors

Strings and vectors of either fixed or variable length are the most elementary search spaces. They are the most frequently used genotype structures. Vectors allow us to represent an ordered list of decision variables and are the standard genotypes for the majority of optimization problems (compare Sects. 3.2 and 3.3). Strings are appropriate for sequences of characters or patterns. Consequently, strings are suited for problems where the objects modeled are "text", "characters", or "patterns".

For strings as well as vectors, we can use standard local search and recombination operators (Sects. 4.3.2 and 4.3.3) that are mostly based on the Hamming metric (2.7) or binary Hamming metric (2.8).

4.3.5.2 Coordinates/points

To represent locations in a geometric space, coordinates can be used. Coordinates can be either integer or continuous. Common examples are locations of cities or other spots on 2-dimensional grids. Coordinates are appropriate for problems that work on "sites", "positions", or "locations".

We can use standard local and recombination operators for continuous variables and integers, respectively. For coordinates, often the Euclidean metric is used to measure the similarity of solutions.

4.3.5.3 Graphs

Graphs allow us to represent relationships between arbitrary objects. Usually, the structure of a graph is described by listing its edges. An edge represents a relationship between a pair of objects. Given a graph with n nodes (objects), there are $n(n-1)/2$ possible edges. Using graphs is appropriate for problems that seek a "network", "circuit", or "relationship".

Common genotypes for graphs are lists of edges indicating which edges are used. Often, the characteristic vector representation (Sect. 8.1.2) or variants of it are used to represent graph structures. Standard search operators for the characteristic vector representation are based on the Hamming metric (2.8) as the distance between two graphs can be calculated as the number of different edges. Standard search operators can be used if there are no additional constraints.

4.3.5.4 Subsets

Subsets represent selections from a set of objects. Given n different objects, the number of subsets having exactly k elements is equal to $\binom{n}{k}$. Thus, the number of possible subsets can be calculated as $\sum_{k=0}^{n} \binom{n}{k} = 2^n$. For subsets, the order of the objects does not matter. Therefore, the two example subsets $\{1,3,5\}$ and $\{3,5,1\}$ represent the same phenotype solution. Local search operators that can be applied directly to subsets often either modify the objects in the subset, or increase/reduce the number of objects in one subset. Recombination operators that are directly applied to subsets are more sophisticated as no standard operators can be used. We refer to Falkenauer (1998) for detailed information on the design of search operators for subsets. Subsets are often used for problems that seek a "cluster", "collection", "partition", "group", "packaging", or "selection".

Given n different objects, a subset of fixed size k can be represented using an integer vector x of length k, where the x_i indicate the selected objects and $x_i \neq x_j$, for $i \neq j$ and $i, j \in [1,k]$. Then, standard local search operators can be applied if we assume that each of the k selected objects is unique. The application of recombination operators is more demanding as each subset is represented by $k!$ different genotypes (integer vectors) and the distances between the $k!$ different genotypes that represent the same subset are large (Choi and Moon, 2008). Recombination operators must be designed such that the distances between offspring and parents are smaller than the distances between parents (4.1) and the recombination of two genotypes that represent the same subset always results in the same subset. For guidelines on the design of appropriate recombination operators and examples, we refer the interested reader to Choi and Moon (2003) and Choi and Moon (2008).

4.3.5.5 Permutations

A large variety of modern heuristics have been developed for permutation problems as many such problems are of practical relevance but NP-hard. Permutations are orderings of items. Relevant for permutations is the order of the objects. The number of permutations on a set of n elements is given by $n!$. 1-2-3 and 1-3-2 are two examples of permutations of three integer numbers $x \in \{1,2,3\}$. The TSP is a prominent example of a permutation problem (Sect. 3.3.2). Permutations are commonly used for problems that seek an "arrangement", "tour", "ordering", or "sequence".

The design of appropriate search operators for permutations is demanding. In most approaches, permutations are encoded using an integer genotype vector of length n, where each decision variable x_i indicates an object and has a unique value ($x_i \neq x_j$ for $i \neq j$ and $i, j \in \{1, \ldots, l\}$). Standard recombination and mutation operators applied to such genotypes fail since the resulting solutions usually represent no permutations. Therefore, in the literature a variety of different permutation-specific variation operators have been developed. They are either based on the absolute or relative ordering of the objects in a permutation. When using the absolute ordering of objects in a permutation as distance metric, two solutions are similar to each other if the objects have the same position in the two solutions ($x_i^1 = x_i^2$). For example, 1-2-3-4 and 2-3-4-1 have a maximum absolute distance of $d = 4$, as the two solutions have no common absolute positions. In contrast, when using relative ordering, two solutions are similar if the sequence of objects is similar for the two solutions. For example, 1-2-3-4 and 2-3-4-1 have distance $d = 1$ as the two permutations are shifted by one position. Based on the metric used (relative versus absolute ordering), a large variety of different recombination and local search operators have been developed. Examples are the *order crossover* (Davis, 1985), *partially mapped crossover* (Goldberg and Lingle, Jr., 1985), *cycle crossover* (Oliver et al, 1987), *generalized order crossover* (Bierwirth, 1995), or *precedence preservative crossover* (Bierwirth et al, 1996). For more information on the design of such permutation-specific variation operators, we refer to Whitley (1997), Mattfeld (1996), and Choi and Moon (2008).

4.3.5.6 Trees

Trees are used to describe hierarchical relationships between objects. Trees are a specialized variant of graphs where only one path exists between each pair of nodes. As standard search operators cannot be applied to tree structures, we either need to define problem-specific search operators that are directly applied to trees or additional genotype-phenotype mappings that map each tree to simpler genotypes where standard variation operators can be applied.

We can distinguish between trees of fixed and variable size. For trees of fixed size, search operators are presented in Sect. 8.3 and appropriate genotypes and genotype-phenotype mappings in Sect. 8.4. Search operators for tree structures of variable size are at the core of *genetic programming* and are discussed in Chap. 7. Further information about appropriate search operators for trees of variable size can be found in Koza (1992) and Banzhaf et al (1997).

4.4 Fitness Function

Modern heuristics use a *fitness function* to compare the quality of solutions. The fitness of a solution is its quality "seen" by the optimization method. It is based on the *objective function* (which is often also called *evaluation function*) and often both are equivalent to each other. However, sometimes modern heuristics modify the objective function and use the resulting fitness function to compare the quality of different solutions. In general, the objective function is based on the problem and model formulation whereas the fitness function measures the quality of solutions from the perspective of modern heuristics. In the following paragraphs, we discuss some relevant aspects of objective and fitness functions.

The objective function is usually defined for all problem solutions and allows us to compare the quality of different solutions. The objective function is based on the problem model and is a mapping from the set of possible candidate solutions to a set of objective values. We can distinguish between two types of objective functions: *ordinal* and *numerical*. Ordinal objective functions denote the position of a solution in an ordered sequence. With $Y = \{1, \ldots, |\Phi|\}$,

$$f(x) : \Phi \rightarrow Y$$

indicates the position of a solution in the sequence. Ordinal objective functions allow us to order the solutions with respect to their quality (best, second best, ..., worst) but give us no information on their absolute quality. For example, ordinal objective functions are commonly used when human experts are involved in the evaluation of solutions. For humans, it is usually easier to establish a ranking of potential solutions instead of assigning absolute quality values.

Numerical objective functions assign a real-valued objective value to all possible solutions which indicates the quality of a solution:

$$f(x) : \Phi \rightarrow \mathbb{R}.$$

Based on the objective value, we can order solutions with respect to their quality. Numerical objective functions are common for many technical or mathematical optimization problems, where the goal is to minimize or maximize some quality measurement (e.g. cost or profit).

The objective function is formulated during model construction (Sect. 2.1.3) and depends on the problem definition. Usually, we have different possibilities for formulating the objective function. When defining an objective function, we should ensure that optimal solutions (solutions that meet the objective completely) have the best evaluation. Therefore, solutions that have lower quality than the optimal solution also have a lower objective value. Furthermore, solutions of similar quality should have similar objective values. Finally, objective functions should assign objective values to the solutions in such a way that guided search methods can easily find the optimal solution and are "guided" towards the optimal solutions by the structure of the fitness landscape. Therefore, objective functions should make problems either straightforward (Sect. 2.4.2.1) if local search operators are used, or decomposable (Sect. 2.4.3) if recombination search operators are used, or both. Therefore, the dissimilarity between phenotype solutions (measured by the problem metric) should be positively correlated with the difference in their objective values.

We want to give two examples of a bad design of objective functions. In the first example, we have a search space X of size n. The objective function assigns the highest objective value to the best solution ($f(\max_X x) = n$) and randomly assigns an objective value $\{1,\ldots,n-1\}$ to the other $n-1$ solutions $x \in X - \{\max_X x\}$. This problem is equivalent to a NIH problem and is closed under permutation (Sect. 3.4.4). Therefore, guided search methods have large problems finding the optimal solution. On average, they cannot perform better than random search as they cannot exploit any information that leads them in the direction of optimal solutions.

The second example is the *satisfiability (SAT) problem* (Cook, 1971; Garey and Johnson, 1979; Gu et al, 1996). An instance of the SAT problem is a Boolean formula with three components:

- A set of n variables x_i, where $i \in \{1,\ldots,n\}$.
- A set of literals. A literal is a variable or a negation of a variable.
- A set of m distinct clauses $\{C_1, C_2, \ldots, C_m\}$. Each clause consists only of literals combined by logical *or* operators ($\vee$).

The SAT is a decision problem and its objective is to determine whether there exists an assignment of values to the n variables such that the conjunctive normal form $C_1 \wedge C_2 \wedge \cdots \wedge C_m$ becomes satisfiable, where $\wedge$ is the logical *and* operator. The most natural objective function for the SAT problem assigns a zero objective value to all solutions that do not satisfy the given compound Boolean statement and a one to all solutions that satisfy the statement. However, such an objective function results in large problems for guided search methods as the resulting problem is a needle-in-a-haystack problem and it is not possible to extract information from the search history to guide a search method in the direction of an optimal solution. A more appropriate fitness function would use a measure for the quality of a solution, for example the number of satisfied clauses. This would allow modern heuristics to estimate how feasible a solution is (how many clauses are correct) and distinguish between solutions based on their "degree" of inadmissibility.

From the second example, we can learn two important lessons for the design of fitness functions. First, fitness landscapes with large plateaus like in the NIH problem can be made easier for guided search methods if we modify the objective function and consider the objective value of neighboring solutions for calculating a solution's fitness. For example, this can be achieved by *smoothing* the fitness landscape. Smoothing algorithms are commonly used in statistics to allow us to recognize relevant trends in data (Simonoff, 1996). We can use this concept for optimization and calculate the fitness value of a solution considering the neighboring solutions. Smoothing has the drawback that we need additional fitness evaluations as we must know for each evaluation the fitness values of neighboring solutions. An example of smoothing is calculating the fitness of a solution as the average of the weighted neighboring solutions (the weight decreases with increasing distance). Thus, we can transform plateaus or NIH-landscapes into fitness landscapes that guide local search methods towards optimal solutions (see Fig. 4.3 for an example).

Second, when dealing with constrained problems, we have to define fitness values for solutions that are infeasible (often known as *penalty functions*). Defining fitness

Fig. 4.3 Smoothing

values for infeasible solutions is necessary if not all infeasible solutions can be excluded from the search space, for example by a proper definition of the genotype space or by search operators that do not generate infeasible solutions. As we have seen in the SAT example, we sometimes cannot systematically exclude all infeasible solutions from the solution space but also have to assign a fitness value to infeasible solutions. The proper design of fitness functions for constrained problems is demanding as often the optimal solutions are at the edge of the feasible search space (Gottlieb, 1999) and we have to design fitness functions for infeasible solutions such that the resulting problem is either straightforward (when using local search operators), decomposable (when using recombination operators), or both. For details on the design of appropriate fitness functions for constrained problems we refer to the literature (Smith and Coit, 1997; Gottlieb, 1999; Coello Coello, 1999; Michalewicz et al, 1999)

As modern heuristics usually need a large number of solution evaluations, the calculation of the fitness values must be fast. Problems where the calculation is very time-consuming often cannot be solved efficiently by modern heuristics. In many real-world problems, there is a trade-off between speed and accuracy of the evaluations (Handa, 2006). We can get rough approximations of a solution's quality after a short time but often need more time to calculate the exact quality. As the differences in the objective values of solutions at the beginning of optimization runs are usually large, we do not need accurate quality estimations at early stages of a run. With increasing run time, modern heuristics find better and more similar solutions and accuracy becomes more important. When designing the fitness function for real-world problems where solution evaluation is time-consuming, users have to take these aspects into account.

4.5 Initialization

The proper choice of an initialization method is important for the design of effective and efficient modern heuristics. An initialization method provides modern heuristics with some initial solutions from which the search starts. We can distinguish between initialization methods that generate either only one solution or a population of solutions. If we want to use local search approaches, one solution may be sufficient; for recombination-based approaches, we need a larger set of individuals since we can only recombine properties from different solutions if the number of available solutions is large enough.

Proper initialization strategies for modern heuristics depend on the available knowledge about the problem. If no problem knowledge exists and we have no information about the properties of high-quality or low-quality solutions, we recommend the creation of random solutions that are uniformly distributed in the search space. Then, all solutions are selected with equal probability. Therefore, we have to ensure that the initialization method is unbiased and does not favor some types of solutions. This can sometimes be difficult, especially if direct representations are used, as a possible bias of phenotypes is often difficult to recognize. The situation is easier for standard genotypes like continuous, integer, or binary decision variables. For such types of solutions, we can more easily recognize a bias and create unbiased, random solutions.

The situation is different if we have some kind of knowledge about the problem or different types of solutions. This can be, for example, knowledge about the range of variables, dependencies between different variables, or properties of high-quality or low-quality solutions. In general, initial solutions should be created such that such knowledge is considered and initial solutions are biased towards high-quality solutions. For the construction of appropriate initial solutions, construction heuristics (Sect. 3.4.1) are often used. Construction heuristics are based on existing knowledge about the structure of optimal solutions and are designed to create high-quality solutions. Therefore, construction heuristics, or variants of them, are often a good choice for creating initial solutions if they appropriately consider problem-specific knowledge.

The *diversity* of a population (Holland, 1975) measures the variation in a set of solutions. Diversity is high if the average distance between solutions is large; it is low if the average distance is small. Due to problems with low diversity, we have to be careful when using recombination-based search approaches and biased populations (such as those, for example, created by a construction heuristic). Low diversity leads to problems for recombination-based search because the number of available values of the decision variables that can be recombined to form new solutions is limited and it may happen that optimal solutions cannot be formed by recombination. Low diversity can be the result of a (too) small population or due to biased initial solutions. Because biased initial populations often have low diversity, not all relevant parts of the search space can be explored and recombination-based search fails to find optimal solutions that have a larger distance to the solutions in the biased initial population. In general, biased initial solutions with low diversity and small populations can be problematic for recombination-based search and must be used with care as they limit the ability of modern heuristics to create diversified populations.

Biased initial solutions can be problematic also for local search approaches as a bias usually focuses the search on a specific part of the search space and reduces the probability of finding high-quality solutions that have a large distance to the biased initial solution or population.

Therefore, we have to be careful when designing initialization methods. If possible, initialization should be biased towards high-quality solutions. However, if we have only fuzzy knowledge about the structure of high-quality or low-quality solutions, we should avoid a strong bias as it can mislead heuristic search. Furthermore, biased initial populations should have sufficient diversity to make sure that recombination operators work properly and also solutions that are different from the biased initial population can be found.

Sections 8.3 and 8.5 present illustrative examples of how biased initialization methods influence the performance of modern heuristics. The results show that modern heuristics for the OCST problem can benefit from biasing initial solutions towards minimum spanning trees (MST). However, if optimal solutions have a larger distance to MSTs, a strong initialization bias towards MSTs results in problems for modern heuristics.

Chapter 5
Search Strategies

In the previous chapter, we discussed important elements of modern heuristics. We gave an overview of representations and search operators and discussed various aspects of the design of fitness functions and initialization methods. This section delivers the final design element and describes concepts for controlling the search. Different strategies for controlling search differ in the design and control of the intensification and diversification phases (see Sect. 3.4.3 and Blum and Roli (2003)). It is important for search strategies to balance intensification and diversification during search and to allow search methods to escape from local optima. This is achieved by various diversification techniques based on the representation, search operator, fitness function, initialization, or explicit diversification steps controlled by the search strategy.

In analogy to search operators, we distinguish between two fundamental concepts for heuristic search: local search methods versus recombination-based search methods. The choice between these two different concepts is problem-specific. If the problem at hand has high locality and the distances between solutions correspond to their fitness difference (Sect. 2.4.2), local search methods are the methods of choice. Then, the structure of the search space guides local search methods towards optimal solutions and local search outperforms random search (see the discussion in Sect. 3.4.4). The situation is slightly different for recombination-based search approaches, as these methods show good performance if the problem at hand is decomposable (Sect. 2.4.3). Then, the problem can be solved by decomposing the problem into smaller subproblems, solving these subproblems independently of each other, and determining the overall optimal solution by combining optimal solutions for the subproblems. As many real-world problems have high locality and are decomposable, both types of search methods often show good performance and are able to return high-quality solutions. However, direct comparisons between these two concepts are only meaningful for particular problem instances, and general statements on the superiority of one or other of these basic concepts are unjustified. Which of the two concepts is more appropriate for solving a particular problem instance depends on the specific characteristics of the problem (locality versus decomposability).

F. Rothlauf, *Design of Modern Heuristics*, Natural Computing Series,
DOI 10.1007/978-3-540-72962-4_5, © Springer-Verlag Berlin Heidelberg 2011

Therefore, comparing the performance of different search strategies by studying their performance for a set of predefined test functions and generalizing the results can be problematic (Thierens, 1999), as test functions are usually biased to favor one particular concept. For example, common test functions for recombination-based search strategies are usually decomposable. Representative examples for decomposable problems are the one-max problem (also known as the bit-counting problem), concatenated deceptive traps (Ackley, 1987; Deb and Goldberg, 1993a), or royal road functions (Mitchell et al, 1992; Jansen and Wegener, 2005). For such types of test problems, recombination-based search approaches often show better performance than local search approaches. In contrast, problems that are commonly used as test problems for local search methods often show a high degree of locality (e.g. the corridor model (5.1), sphere model (5.2), or the Rosenbrock function (3.1)) resulting in high performance of local search methods.

This chapter starts with a discussion on how different search strategies ensure diversification in the search. Diversification can be introduced into search by a proper design of a representation or operator, a fitness function, initial solutions, or an explicit control of the search strategy. Section 5.1 gives an overview of representative local search approaches that follow these principles. We describe variable neighborhood search, guided local search, iterated local search, simulated annealing, Tabu search, and evolution strategies. We especially discuss how the different approaches balance diversification and intensification. Section 5.2 focuses on recombination-based search methods and discusses representative approaches like genetic algorithms, estimation of distribution algorithms, and genetic programming. Again, we study intensifying and diversifying elements of the search strategies.

5.1 Local Search Methods

The idea of local search is to iteratively create neighboring solutions. Since such strategies usually consider only one solution, recombination operators are not meaningful. For the design of efficient local search methods it is important to incorporate intensification as well as diversification phases into the search.

Local search methods are also called *trajectory methods* since the search process can be described as a trajectory in the search space. The search space is a result of the interplay between representation and operator. A trajectory depends on the initial solution, the fitness function, the representation/operator combination, and the search strategy used. The behavior and the dynamics of local as well as recombination-based search methods can be described using *Markov processes* and concepts of statistical mechanics (Rudolph, 1996; Vose, 1999; Reeves and Rowe, 2003). In Markov processes, states are used which represent the subsequently generated solutions (or populations of solutions). A search step transforms one state into a following state. The behavior of search algorithms can be analyzed by studying possible sequences of states and their corresponding probabilities. The transition matrix describes how states depend on each other and depends on the initial solution, fitness function, representation/operator combination, and search strategy.

Existing local as well as recombination-based search strategies mainly differ in how they control diversification and intensification. Diversification is usually achieved by applying variation operators or making larger modifications of solutions. Intensification steps use the fitness of solutions to control search and usually ensure that the search moves in the direction of solutions with higher fitness. The goal is to find a trajectory that overcomes local optima by using diversification and ends in a global optimal solution.

In greedy search approaches (like the best-first search strategy discussed in Sect. 3.3.2), intensification is maximal as in each search step the neighboring solution with highest fitness is chosen. No diversification is possible and the search stops at the nearest local optimum. Therefore, greedy search finds the global optimum if we have an unimodal problem where only one local optimum exists. However, as problems usually have a larger number of local optima, the probability of finding the global optimum using greedy search is low.

Based on the design elements of modern heuristics, there are different strategies to introduce diversification into the search and to escape from local optima:

- Representation and search operator: Choosing a combination of representation and search operators is equivalent to defining a metric on the search space and defines which solutions are neighbors. By using different types of neighborhoods, it is possible to escape from local optima and explore larger areas of the search space. Different neighborhoods can be the result of different genotype-phenotype mappings or search operators applied during search. Standard examples for local search approaches that use modifications of representations or operators to diversify the search are variable neighborhood search (Hansen and Mladenović, 2001), problem space search (Storer et al, 1992) (see also Sect. 6.2.1), the rollout algorithm (Bertsekas et al, 1997), and the pilot method (Duin and Voß, 1999).
- Fitness function: The fitness function measures the quality of solutions. Modifying the fitness function has the same effect as changing the representation as it assigns different fitness values to the problem solutions. Therefore, variations and modifications of the fitness function lead to increased diversification in local search approaches. A common example is guided local search (Voudouris, 1997; Balas and Vazacopoulos, 1998) which systematically changes the fitness function with respect to the progress of search.
- Initial solution: As the search trajectory depends on the choice of the initial solution (for example, greedy search always finds the nearest local optimum), we can introduce diversification by performing repeated runs of search heuristics using different initial solutions. Such multi-start search approaches allow us to explore a larger area of the search space and lead to higher diversification. Variants of multi-start approaches include iterated descent (Baum, 1986a,b), large-step Markov chains (Martin et al, 1991), iterated Lin-Kernighan (Johnson, 1990), chained local optimization (Martin and Otto, 1996), and iterated local search (Lourenco et al, 2001).

- Search strategy: An important element of modern heuristics are intensification steps like those performed in local search that push the search towards high-quality solutions. To avoid "pure" local search ending in the nearest local optimum, diversification steps are necessary. The search strategy can control the sequence of diversification and intensification steps. Diversification steps that do not move towards solutions with higher quality can either be the results of random, larger, search steps or based on information gained in previous search steps. Examples of search strategies that use a controlled number of search steps towards solutions of lower quality to increase diversity are simulated annealing (Aarts and van Laarhoven, 1985; van Laarhoven and Aarts, 1988) (see Sect. 3.4.3), threshold accepting (Dueck and Scheuer, 1990), or stochastic local search (Gu, 1992; Selman et al, 1992; Hoos and Stützle, 2004). Representative examples of search strategies that consider previous search steps for diversification are tabu search (Glover, 1986; Glover and Laguna, 1997) or adaptive memory programming (Taillard et al, 2001).

The following paragraphs discuss the functionality and properties of selected local search strategies. The different examples illustrate the four different approaches (representation/operator, fitness function, initial solution, search strategy) to introduce diversity into search. The section ends with a discussion of evolution strategies which are representative examples of local search approaches that use a population of solutions. In principle, all local search concepts that have been developed for a single solution can be extended to use a population of solutions.

5.1.1 Variable Neighborhood Search

Variable Neighborhood Search (VNS) (Mladenovic and Hansen, 1997; Hansen and Mladenović, 2001) combines local search strategies with dynamic neighborhood structures that are changed subject to the progress made during search. VNS is based on the following observations (Hansen and Mladenović, 2003):

- A local minimum with respect to a neighborhood structure is not necessarily a local optimum with respect to a different neighborhood (see also Sects. 2.3.2 and 4.2.2). The neighborhood structure of the search space depends on the metric used and is different for different search operators and representations. This observation goes back to earlier work (Liepins and Vose, 1990; Jones, 1995b,a) which found that different types of operators result in different fitness landscapes.
- A global minimum is a global minimum with respect to all possible neighborhood structures. Different neighborhood structures only result in different similarity definitions but do not change the fitness of the solutions. Therefore, the global optimum is independent of the search operators used and remains the global optimum for all possible metrics.
- Hansen and Mladenović (2003) conjecture that in many real-world problems, local optima with respect to different neighborhood structures have low distance to

each other and local optima have some properties that are also relevant for the global optimum. This observation is related to the decomposability of problems which is relevant for recombination-based search. Local optima are not randomly distributed in the search space but local optima already contain some optimal solutions to subproblems. Therefore, as they share common properties, the average distance between local optima is low.

Figure 5.1 illustrates the basic idea of VNS. The goal is to repeatedly perform a local search using different neighborhoods N. The global optimum x^* remains the global optimum with respect to all possible neighborhoods. However, as different neighborhoods result in different neighbors, x can be a local optimum with respect to neighborhood N_1 but it is not necessarily a local optimum with respect to N_2. Thus, performing a local search starting from x and using N_2 can find the global optimum.

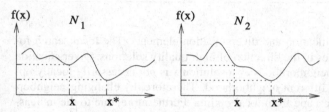

Fig. 5.1 Changing the neighborhood from N_1 to N_2 allows local search to find the global optimum

The functionality of VNS is described in Algorithm 2. During initialization, a set of k different neighborhoods is defined. A neighborhood function $N(x) : X \to 2^X$ (2.9) describes which solutions are neighbors of $x \in X$. It is based on a metric and its cardinality $|N(x)|$ is the (average) number of neighbors of $x \in X$. For VNS, it is important to choose neighborhoods with an appropriate cardinality. At the extreme, if the search space is fully connected and the cardinality of a neighborhood is similar to the problem size (for example $|N(x)| = |X| - 1$), all solutions are neighboring solutions and can be reached in one search step. Then, guided search is not possible any more as we have no meaningful metric to measure similarities between solutions (we have a trivial topology, see Sect. 2.3.1). In principle, the set of k different neighborhoods can be arbitrarily chosen, but often a sequence $|N_1| < |N_2| < \cdots < |N_{k_{max}}|$ of neighborhoods with increasing cardinality is used. VNS iteratively performs a "shaking phase" and a local search phase. During shaking, a solution x' in the kth neighborhood of the current solution x is randomly selected. x' is generated at random to avoid cycling and to lead local search towards a new local optimum different from x. During local search, we start with x' and perform a local search until a local optimum x'' is found. If x'' is better than x, it replaces x and the algorithm starts anew from this solution using the first neighborhood N_1. Otherwise, we continue with shaking.

Algorithm 2 Variable Neighborhood Search

Select a set of neighborhood structures $N_k, k \in \{1, \ldots, k_{max}\}$
Create initial solution x
while termination criterion is not met **do**
   $k = 1$
   while $k < k_{max}$ **do**
      Shaking: choose a random neighbor $x' \in N_k(x)$
      Local search: perform a local search starting with x' and return x'' as the local optimum
      with respect to N_k
      if $f(x'') < f(x)$ (minimization problem) **then**
         $x = x''$
         $k = 1$
      else
         $k = k + 1$
      end if
   end while
end while

VNS contains intensification and diversification elements. The local search focuses search as it searches in the direction of high-quality solutions. Diversification is a result of changing neighborhoods as a solution x is not necessarily locally optimal with respect to a different neighborhood. Therefore, by changing neighborhoods, VNS can easily escape from local optima. Furthermore, due to the increasing cardinality of the neighborhoods (the neighborhoods are ordered with respect to their cardinality), diversification gets stronger as the shaking steps can choose from a larger set of solutions and local search covers a larger area of the search space (the *basin of attraction* increases).

Although in the last few years VNS has become quite popular and many publications have shown successful applications (for an overview see Hansen and Mladenović (2003)), the underlying ideas are older and more general. The goal is to introduce diversification into modern heuristics by changing the metric of the problem with respect to the progress that is made during search. A change in the metric can be achieved either by using different search operators or a different genotype-phenotype mapping. Both lead to different metrics and neighborhoods. Early ideas on varying the representation (adaptive representations) with respect to the search progress go back to Holland (1975). First implementations have been presented by Grefenstette et al (1985), Shaefer (1987), Schraudolph and Belew (1992), and Storer et al (1992) (see also the discussion in Sect. 6.2.1). Other examples are approaches that use additional transformations (Sebald and Chellapilla, 1998b), a set of pre-selected representations (Sebald and Chellapilla, 1998a), or multiple and evolvable representations (Liepins and Vose, 1990; Schnier, 1998; Schnier and Yao, 2000).

5.1.2 Guided Local Search

A fitness landscape is the result of the interplay between a metric that defines similarities between solutions and a fitness function that assigns a fitness value to each
solution. VNS uses modifications of the metric to create different fitness landscapes
and to introduce diversification into the search process. *Guided local search* (GLS)
(Voudouris and Tsang, 1995; Voudouris, 1997) uses a similar principle and dynamically changes the fitness landscape subject to the progress that is made during the
search. In GLS, the neighborhood structure remains constant. Instead, it dynamically modifies the fitness of solutions near local optima so that local search can
escape from local optima.

GLS considers problem-specific knowledge by using the concept of *solution features*. A solution feature can be any property or characteristics that can be used to
distinguish high-quality from low-quality solutions. Examples of solution features
are edges used in a tree or graph, city pairs (for the TSP), or the number of unsatis-
fied clauses (for the SAT problem; see p. 126). The indicator function $I_i(x)$ indicates
whether a solution feature $i \in \{1, \ldots, M\}$ is present in solution x. For $I_i(x) = 1$, solution feature i is present in x, for $I_i(x) = 0$ it is not present. GLS modifies the fitness
function f such that the fitness of solutions with solution features that exist in many
local optimal solutions is reduced. For a minimization problem, $f(x)$ is modified to
yield a new fitness function

$$f'(x) = f(x) + \lambda \sum_{i=1}^{M} p_i I_i(x),$$

with the *regularization parameter* λ and the *penalty parameters* p_i. The p_i are ini-
tialized as $p_i = 0$. M denotes the number of solution features. λ weights the impact
of the solution features on the original fitness function f and p_i balances the impact
of solution features of different importance.

Algorithm 3 describes the functionality of GLS. It starts from a random solution
x_0 and performs a local search returning the local optimum x_1. To escape the local
optimum, a penalty is added to the fitness function f such that the resulting fitness
function h allows local search to escape. The strength of the penalty depends on the
utility u_i which is calculated for all solution features $i \in \{1, \ldots, M\}$ as

$$u_i(x_1) = I_i(x_1) \times c_i/(1 + p_i),$$

where c_i is the cost of solution feature i. The c_i are problem-specific and usually
remain unchanged during search. They are determined by the user and describe
the relative importance of the solution features. Examples of c_i are the weights of
edges (graph or tree problems) or the city-pair distances (TSP). The function $u_i(x)$
is unequal to zero for all solution features that are present in x. After calculating
the utilities, the penalty parameters p_i are increased for those solution features i that
yield the highest utility value. After modifying the fitness function, we start a new

local search from x_1 using the modified fitness function h. Search continues until a termination criterion is met.

Algorithm 3 Guided Local Search

$k = 0$
Create initial solution x_0
for $i = 1$ to M **do**
    $p_i = 0$
end for
while termination criterion is not met **do**
    $h = f + \lambda \sum_{i=1}^{M} p_i I_i$
    perform a local search using fitness function h starting with x_k and return the local optimum
    x_{k+1}
    for $i = 1$ to M **do**
        $u_i(x_{k+1}) = I_i(x_{k+1}) \times c_i/(1 + p_i)$
    end for
    for all i where u_i is maximum **do**
        $p_i = p_i + 1$
    end for
    $k = k + 1$
end while
return x^* (best solution found so far according to f)

The utility function u penalizes solution features i with high cost c_i and allows us to consider problem-specific knowledge by choosing appropriate values for c_i. The presence of a solution feature with high cost leads to a high fitness value of the corresponding solution allowing local search to escape from this local optimum. Figure 5.2 illustrates the idea of changing the fitness of local optima. By modifying the fitness function $f(x)$ and adding a penalty to $f(x)$, $h(x)$ assigns a lower fitness to x. Thus, local search can leave x and is able to find the global optimal solution x^*.

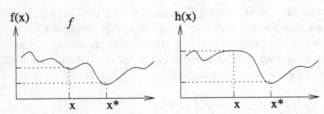

Fig. 5.2 Changing the fitness function from f to h allows local search to find the global optimum

In GLS, diversification is a result of the systematic modification of the fitness function f. The intensity of diversification is controlled by the parameter λ. Large values of λ lead to local search steps that find solution features that were not present in previous solutions. However, if λ is too large, GLS behaves like random search and randomly moves through the search space. Problems can also occur if λ is too low as no information about previous search steps can be considered for the fitness function $h(x)$ and local search repeatedly finds the same local optimum (Mills and Tsang, 2000). In general, the setting of λ is problem-specific and must be done with care.

Examples of the successful application of GLS are TSPs (Voudouris and Tsang, 1999), bin-packing problems (Faroe et al, 2003b), VLSI design problems (Faroe et al, 2003a), and SAT problems (Mills and Tsang, 2000; Zhang, 2004).

5.1.3 Iterated Local Search

Heuristics that use only intensification steps (like local search) are often able to quickly find a local optimal solution but unfortunately cannot leave a local optimum again. A straightforward way to introduce diversification is to perform sequential local search runs using different initial solutions. Such approaches are commonly known as *multi-start approaches*. The simplest variants of multi-start approaches iteratively generate random solutions and perform local search runs starting from those randomly generated solutions. Thus, we have distinct diversification phases and can explore larger areas of the search space. Search strategies that randomly generate initial solutions and perform a local search are also called *multi-start descent* search methods.

However, to randomly create an initial solution and perform a local search often results in low solution quality as the complete search space is uniformly searched and search cannot focus on promising areas of the search space. *Iterated local search* (ILS) (Martin et al, 1991; Stützle, 1999; Lourenco et al, 2001) is an approach to connect the unrelated local search phases as it creates initial solutions not randomly but based on solutions found in previous local search runs. Therefore, it is based on the same observations as VNS which assumes that local optima are not uniformly distributed in the search space but similar to each other (Sect. 5.1.1, p. 134).

Algorithm 4 outlines the basic functionality of ILS. Relevant design criteria for ILS are the modification of x and the acceptance criterion. If the perturbation steps are too small, the following local search cannot escape from a local optimum and again finds the same local optimum. If perturbation is too strong, ILS shows the same behavior as multi-start descent search methods. The modification step as well as the acceptance criterion can depend on the search history.

Algorithm 4 Iterated Local Search

Create initial solution x_0
Perform a local search starting with x_0 and return the local optimum x
while termination criterion is not met **do**
  Modification: perturb x and return x'
  Perform a local search starting with x' and return the local optimum x''
  Acceptance Criterion: decide whether to continue with x or with x''
end while

In ILS, diversification is controlled by the perturbation of the solution (which is problem-specific) and the acceptance criterion. Larger perturbations and continuing search with x'' lead to stronger diversification. Continuing search with x intensifies search as a previously used initial solution is re-used.

A similar concept is used by *greedy randomized adaptive search* (GRASP))Feo et al, 1994; Feo and Resende, 1995). Like in ILS, each GRASP iteration consists of two phases: construction and local search. The construction phase builds a feasible solution, whose neighborhood is investigated until a local minimum is found during the local search phase.

5.1.4 Simulated Annealing and Tabu Search

The previous examples illustrated how modern heuristics can make diversification steps by modifying the representation/search operator, fitness function, or initial solutions. Simulated annealing (SA) is a representative example of a modern heuristic where the search strategy used explicitly defines intensification and diversification phases/steps. The functionality and properties of SA are discussed in detail in Sect. 3.4.3 (pp. 94-97). Its functionality is outlined in Algorithm 1, p. 96.

SA is a combination between a random walk through the search space and local search. Diversification of search is a result of the random walk process and intensification is due to the local search steps. The amount of intensification and diversification is controlled by the parameter T (see Fig. 3.21). With lower temperature T, intensification becomes stronger as solutions with lower fitness are accepted with lower probability. The cooling schedule which determines how T is changed during an SA run is designed such that at the beginning of the search diversification is high whereas at the end of a run pure local search steps are used to find a local optimum.

Tabu search (TS) (Glover, 1977, 1986; Glover and Laguna, 1997) is a popular modern heuristic. To diversify search and to escape from local optima, TS uses a list of previously visited solutions. A simple TS strategy combines such a *short term memory* (implemented as a *tabu list*) with local search mechanisms. New solutions that are created by local search are added to this list and older solutions are removed after some search steps. Furthermore, local search can only create new solutions that do not exist in the tabu list T. To avoid memory problems, usually the length of the tabu list is limited and older solutions are removed. Algorithm 5 outlines the basic functionality of a simple TS. It uses a greedy search, which evaluates all neighboring solutions $N(x)$ of x and continues with a solution x' that is not contained in T and has maximum fitness. Instead of removing the oldest element x_{oldest} from T, also other strategies for updating the tabu list are possible.

The purpose of the tabu list is to allow local search to escape from local optima. By prohibiting previously found solutions, new solutions must be explored. Figure 5.3 illustrates how the tabu list T contains all solutions of high fitness that are in the neighborhood of x. Therefore, new solutions with lower fitness are created and local search is able to find the global optimum.

Algorithm 5 Simple Tabu Search

Create initial solution x
Create empty tabu list T
while termination criterion is not met **do**
$\quad x' = \max\limits_{N(x)-T} f(x)$ (Choose x' as neighboring solution of x with highest fitness that is not con-
$\quad$ tained in T)
$\quad T = T + \{x'\}$
$\quad$ **if** $|T| > l$ **then**
$\quad\quad$ Remove oldest element x_{oldest} from T
$\quad$ **end if**
$\quad x := x'$
end while

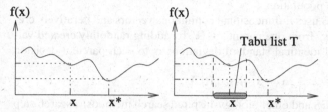

Fig. 5.3 Removing solutions $x \in T$ from the search space allows TS to escape from a local optimum

Diversification can be controlled by the length of the tabu list. Low values lead to stronger intensification as the memory of the search process is smaller. In many applications, the length of the tabu list is dynamically changed and depends on the progress made during search (Glover, 1990; Taillard, 1991; Battiti and Tecchiolli, 1994; Battiti and Protasi, 2001). Such approaches are also known as *reactive Tabu search*.

More advanced TS approaches do not store complete solutions in the tabu list but *solution attributes*. Similarly to the solution features used in GLS, solution attributes can be not only components of solutions but also search steps or differences between solutions. When using different solution attributes, we obtain a set of tabu criteria (called *tabu conditions*) which are used to filter the neighborhood of a solution. To overcome the problem that single tabu conditions prohibit the creation of appropriate solutions (each attribute prohibits the creation of a set of solutions), *aspiration criteria* are used which overwrite tabu conditions and allow search steps to create solutions where some tabu conditions are present. A commonly used aspiration criterion selects solutions with higher fitness than the current best one. For more information on TS and on application examples, we refer to the literature (Glover and Laguna, 1997; Gendreau, 2003; Voß et al, 1999; Osman and Kelly, 1996).

5.1.5 Evolution Strategy

Evolution strategies (ES) are local search methods for continuous search spaces. They were developed by Rechenberg and Schwefel in the 1960s at the Technical University of Berlin (Rechenberg, 1973a,b) and first applications were experimental and dealt with hydrodynamical problems like shape optimization of a bent pipe, drag minimization of a joint plate (Rechenberg, 1965), or structure optimization of a two-phase flashing nozzle (Schwefel, 1968). There are two different types of ES: Simple strategies that use only one individual (like the local search strategies discussed in the previous paragraphs) and advanced strategies that use a set of solutions which is called a population. The use of a population allows ES to exchange information between solutions in the population.

The simple $(1+1)$-*ES* uses n-dimensional continuous vectors and iteratively creates one offspring $x' \in \mathbb{R}^n$ from one parent $x \in \mathbb{R}^n$ by adding randomly created values with zero mean and identical standard deviations σ to each parental decision variable x_i.

$$x'_i = x_i + \sigma \mathcal{N}_i(0,1),$$

where $i \in \{1, \ldots, n\}$. In ES and other biology-inspired search methods, a search step is usually denoted as a *mutation*. $\mathcal{N}(0,1)$ is a normally distributed one-dimensional random variable with expectation zero and standard deviation one. $\mathcal{N}_i(0,1)$ indicates that the random variable is sampled anew for each possible value of the counter i. In the $(1+1)$-ES, the resulting individual is evaluated and compared to the original solution. The better one survives to be used for the creation of the next solution. Algorithm 6 summarizes the basic functionality (maximization problem).

Algorithm 6 $(1+1)$-Evolution Strategy

Create initial solution x
while termination criterion is not met **do**
    for $i = 1$ to n **do**
        $x'_i = x_i + \sigma \mathcal{N}_i(0,1)$
    end for
    if $f(x') \geq f(x)$ **then**
        $x := x'$
    end if
    Update σ (e.g. according to 1/5-rule)
end while

For the $(1+1)$-ES, theoretical convergence models for two simple problems, the *sphere model* and the *corridor model*, exist (Rechenberg, 1973a). Both problems are standard test problems for continuous local search methods. The corridor model is representative of the situation where the current solution x has a large distance to the optimum x^*:

$$f_{corridor}(x) = c_0 + c_1 x_1 \quad \forall i \in \{2, \ldots, n\} : -b/2 \leq x_i \leq b/2 \qquad (5.1)$$

and the sphere model describes the situation near the optimal solution x^*:

$$f_{sphere}(x) = c_0 + c_1 \sum_{i=1}^{n} (x_i - x_i^*)^2, \tag{5.2}$$

where c_0, c_1, and b are problem parameters.

$\zeta(t)$, where t indicates the number of search steps, is defined as the ratio of successful search steps that find a better solution to all search steps (for example, averaged over the last $10n$ search steps). For the corridor and sphere models, a ratio $\zeta(t) = \frac{1}{5}$ leads to maximal convergence speed (Rechenberg, 1973a). Therefore, if $\zeta(t)$ is greater than $\frac{1}{5}$, the standard deviation should be increased, if it is smaller, it should be decreased (Rechenberg, 1973a, p. 123). For sphere and corridor models this $1/5$-rule results in fastest convergence. However, sometimes the probability of success cannot exceed $1/5$. For problems where the fitness function has discontinuous first partial derivatives, or at the edge of the allowed search space, the $1/5$-success rule does not work properly. Especially in the latter case, the success rule progressively forces the sequence of new solutions nearer to the boundary and continuously reduces the step length without approaching the optimum with comparable accuracy (Schwefel, 1995).

In $(1+1)$-ES, the strength of diversification is controlled by the standard deviation σ. With increasing σ, step size increases and new solutions are less similar to the original solutions. If σ is too large, subsequently generated solutions are not related to each other and we obtain a random search behavior. The $1/5$-rule adjusts σ and ensures that it becomes small enough to generate similar solutions. However, in contrast to a fixed σ, the $1/5$-rule does not allow $(1+1)$-ES to escape from local optima. Therefore, the $1/5$-rule is not appropriate for multi-modal problems with more than one local optimum as it focuses the search on promising areas of the search space by reducing the step size. The second intensifying element in $(1+1)$-ES is the selection process which continues with the better solution. This process is equivalent to local search.

Schwefel (1975) introduced the principle of a population into ES and proposed the $(\mu+\lambda)$-ES and the (μ,λ)-ES to overcome problems of the $(1+1)$-ES in escaping from local optima. In population-based ES approaches, a set of μ solutions (usually called *individuals*) forms a parent population. In each ES iteration (called a *generation*) a set of λ new solutions (offspring population) is created. For the $(\mu+\lambda)$-ES, the next parent population is created by choosing the μ best individuals from the union of the parent and offspring population ("+"-selection). In the case of the (μ,λ)-ES, where $\lambda > \mu$, only the best μ solutions from the offspring population are chosen (","-selection). Both population-based ES start with a parent population of μ randomly generated individuals. Usually, each individual consists of an n-dimensional vector $x \in \mathbb{R}^n$ and an m-dimensional vector of standard deviations $\sigma \in \mathbb{R}^m_+$ (Bäck, 1996).

To exchange information between solutions, recombination operators are used in population-based ES as background search operators. Mostly, discrete crossover is used for creating the decision variables x'_i and intermediate crossover is used for the standard deviations σ'_i. For discrete crossover, x'_i is randomly taken from one parent, whereas intermediate crossover creates σ'_i as the arithmetic mean of the parents' standard deviations (see Sect. 4.3).

The main search operator in ES is mutation. It is applied to every individual after recombination:

$$\sigma'_k = \sigma_k \exp(\tau_1 N(0,1) + \tau_2 N_k(0,1)) \quad \forall k = 1, 2, \ldots, m, \qquad (5.3)$$

$$x'_i = x_i + \sigma'_i \mathcal{N}_i(0,1) \quad \forall i = 1, 2, \ldots, n, \qquad (5.4)$$

where τ_1 and τ_2 are method-specific parameters and usually $m = n$. If $m \neq n$, some standard deviations σ'_i are used for more than one decision variable. The standard deviations σ_k are mutated using a multiplicative, logarithmic normally distributed process with the factors τ_1 and τ_2. Then, the decision variables x_i are mutated by using the modified σ'_k. This mutation mechanism enables the ES to evolve its own strategy parameters during the search. The logarithmic normal distribution is motivated as follows. A multiplicative modification process for the standard deviations guarantees positive values for σ and smaller modifications must occur more often than larger ones (Schwefel, 1977). Because the factors τ_1 and τ_2 are robust parameters, Schwefel (1977) suggests setting them as follows: $\tau_1 \propto (\sqrt{2\sqrt{n}})^{-1}$ and $\tau_2 \propto (\sqrt{2n})^{-1}$. Newer investigations indicate that optimal adjustments are in the interval $[0.1, 0.2]$ (Kursawe, 1999; Nissen, 1997). τ_1 and τ_2 can be interpreted as "learning rates" as in artificial neural networks, and experimental results indicate that the search process can be tuned for particular objective functions by modifying τ_1 and τ_2.

One of the major advantages of ES is its ability to incorporate the most important parameters of the strategy, e.g. standard deviations, into the search process. Therefore, optimization not only takes place on object variables, but also on strategy parameters according to the actual local topology of the fitness function. This capability is called *self-adaption*.

In population-based ES, intensification is a result of the selection mechanism which prefers high-quality solutions. Recombination is a background operator which has both diversifying and intensifying character. By recombining two solutions, new solutions are created which lead to a more diversified population. However, especially intermediate crossover leads to reduced diversity during an ES run since the population converges to the mean values. Like for $(1 + 1)$-ES, the main source of diversification is mutation. With larger standard deviations σ_i, diversification gets stronger as the step size increases. The balance between diversification and intensification is maintained by the self-adaptation of the strategy parameters. Solutions with standard deviations that result in high-quality solutions stay in the population, whereas solutions with inappropriate σ_i are removed by the selection process.

$(\mu + \lambda)$-ES and (μ, λ)-ES are examples of local search strategies that use a population of solutions. Although local search is the main search strategy, the existence of a population allows ES to use recombination operators which exchange information between different solutions. For further information and applications of ES we refer to the literature (Schwefel, 1981; Rechenberg, 1994; Schwefel, 1995; Bäck and Schwefel, 1995; Bäck, 1996, 1998; Beyer and Deb, 2001; Beyer and Schwefel, 2002).

5.2 Recombination-Based Search Methods

In contrast to local search approaches which exploit the locality of problems and show good performance for problems with high locality, recombination-based approaches make use of the decomposability of problems and perform well for decomposable problems. Recombination-based search solves decomposable problems by decomposing them into smaller subproblems, solving those smaller subproblems separately, and combining the resulting solutions for the subproblems to form overall solutions (Sect. 2.4.3). Hence, the main search operator used in recombination-based search methods is, of course, recombination.

Recombination operators should be designed such that they properly decompose the problem and combine high-quality sub-solutions in different ways (Goldberg, 1991a; Goldberg et al, 1992). A proper decomposition of problems is the key factor for the design of successful recombination operators. Recombination operators must be able to identify the relevant properties of a solution and transfer these as a whole to offspring solutions. In particular, they must detect between which parts of a solution a meaningful linkage exists and not destroy this linkage when creating an offspring solution. This property of recombination operators is often called *linkage learning* (Harik and Goldberg, 1996; Harik, 1997). If a problem is decomposable, the proper problem decomposition is the most demanding part as usually the smaller subproblems can be solved much more easily than the full problem. However, in reality, most often a problem is not completely separable but there exists still some linkage between different sub-problems. Therefore, usually it is not enough for recombination-based methods to be able to decompose the problem and solve the smaller sub-problems, but usually they must also try different combinations of high-quality sub-solutions to form an overall solution. This process of juxtaposing various high-quality sub-solutions to form different overall solutions is often called *mixing* (Goldberg et al, 1993b; Thierens, 1995; Sastry and Goldberg, 2002).

Recombination operators construct new solutions by recombining the properties of parent solutions. Therefore, recombination-based modern heuristics usually use a population of solutions since when using only single individuals, like in most local search approaches, no properties of different solutions can be recombined to form new solutions. Consequently, the main differences between local and recombination-based search are the use of a recombination operator and a population of solutions.

For local search approaches, we have been able to classify methods with respect to their main source of diversification. In principle, we can use the same mechanisms for recombination-based search methods, too. However, in most implementations of recombination-based search methods, an initial population of diverse solutions is the only source of diversification. Usually, no additional explicit diversification mechanisms based on a systematic modification of representations/search operators or fitness functions are used. Search starts with a diversified population of solutions and intensification mechanisms iteratively intensify the search. Recombination-based search usually stops after the population has almost converged. This means that, at the end of the search, only little diversity in the population exists and all individuals are about the same. During a run, recombination (and most often also local search) operators are applied to parent solutions to create new solutions with similar properties. Usually, these variation operators do not increase the diversity in the population but create new solutions with similar properties.

When the population is almost homogeneous, the only source of diversification are small mutations which usually serve as background noise. Problems occur for recombination-based search if the population is almost converged (either to the correct optimal solution or a sub-optimal solution) but search continues. Then, recombination cannot work properly any more (since it needs a diversified population) and search relies solely on local search steps. However, local search steps are usually small and have only little effect. The situation that a population of solutions converges to a non-optimal solution and cannot escape from this local optimum is often called *premature convergence* (Goldberg, 1989c; Collins and Jefferson, 1991; Eshelman and Schaffer, 1991; Leung et al, 1997). In this situation and if no explicit diversification mechanisms are used, only local search and no recombination-based search is possible any more.

Possible strategies to overcome premature convergence and to keep or re-introduce diversification into a population could be based on the representation/operator, fitness function, initial solution, or search strategy and work in a similar way as described in the previous section on local search methods. A straightforward approach to increase diversity in the initial population is to increase the number of solutions in the population. Other approaches address operator aspects and prevent for example recombination between similar solutions in the population (Eshelman and Schaffer, 1991) or design operators which explicitly increase diversity (Bäck, 1996). Low-locality search operators and representations (Sect. 6.1) are other examples since such operators and representations randomize the search process and thus increase population diversity.

Many of the existing approaches limit the intensification strength of the *selection process*. Selection decides which solutions remain in the population and are used for future search steps. The selection process is equivalent to local search when using only one single solution. In the literature, approaches have been presented that decrease intensification by continuously reducing selection intensity during a run and not focusing only on better solutions but also accepting to some extent worse solutions (Bäck, 1996; Goldberg and Deb, 1991). Other approaches to reduce the intensifying character of the selection process limit the number of similar solutions in the population by either letting a new solution replace only a worse solution similar to itself (De Jong, 1975) or only adding a new solution to a population if it is entirely different (Whitley, 1989).

Finally, a substantial amount of work focuses on *niching methods* ((Deb and Goldberg, 1993b; Hocaoglu and Sanderson, 1995; Mahfoud, 1995; Horn and Goldberg, 1998). Niching methods deal with the simultaneous formation and evolution of different sub-populations in a population. The use of niching methods leads to higher diversification in a population as substantially different sub-populations exist. Niching methods are especially important for multi-objective optimization methods (Deb, 2001; Coello Coello et al, 2007) as such methods should return not only one solution but a set of different Pareto-optimal solutions.

The term *evolutionary algorithm* (EA) denotes optimization methods that are inspired by the principles of evolution and apply recombination, mutation, and selection operators to a population of solutions (for details on the operators, we refer to Sect. 5.2.1). Prominent examples of evolutionary algorithms are genetic algorithms (Sect. 5.2.1), evolution strategies (Sect. 5.1.5), genetic programming (Sect. 5.2.3), and *evolutionary programming* (Fogel et al, 1966; Fogel, 1999). Although the term evolutionary algorithm is commonly used in scientific literature, we do not follow this categorization since it classifies search methods based on their main source of inspiration (algorithms inspired by natural evolution) and not according to the underlying working principles. For example, genetic algorithms and genetic programming use recombination as main search operator, whereas evolution strategies and evolutionary programming use local search. Furthermore, genetic algorithms and *scatter search* (Glover, 1997; Laguna and Martí, 2003) share the same working principles (Glover, 1994), however, since scatter search is not inspired by evolution, it would not be ranked among evolutionary algorithms.

The following paragraphs present the simple genetic algorithm and two variants of it as representative examples of recombina tion-based search. The first variant, estimation of distribution algorithms, uses different types of search operators and the second one, genetic programming, uses solutions of variable length.

5.2.1 Genetic Algorithms

Genetic algorithms (GAs) were introduced by Holland (1975) and imitate basic principles of nature formulated by Darwin (1859) and Mendel (1866). They are (like population-based ES discussed on p. 142) based on three basic principles:

- There is a population of solutions. The properties of a solution are evaluated based on the phenotype, and variation operators are applied to the genotype. Some of the solutions are removed from the population if the population size exceeds an upper limit.
- Variation operators create new solutions with similar properties to existing solutions. In GAs, the main search operator is recombination and mutation serves as background operator. In ES, the situation is reversed.
- High-quality individuals are selected more often for reproduction by a selection process.

To illustrate the basic functionality of GAs, we want to use the standard *simple genetic algorithm* (SGA) popularized by Goldberg (1989c). This basic GA variant is commonly used and well understood and uses crossover as main operator (mutation serves only as background noise). SGAs use a constant population of size N, usually the N individuals $x^i \in \{0,1\}^l$ ($i \in \{1,\dots,N\}$) are binary strings of length l, and recombination operators like uniform or n-point crossover are directly applied to the genotypes. The basic functionality of a SGA is shown in Algorithm 7. After randomly creating and evaluating an initial population, the algorithm iteratively creates new generations by recombining (with probability p_c) the selected highly fit individuals and applying mutation (with probability p_m) to the offspring. The function $random(0,1)$ returns a uniformly distributed value in $[0,1)$.

Algorithm 7 Simple Genetic Algorithm

Create initial population P with N solutions x^i ($i \in \{1,\dots,N\}$)
for $i = 1$ to N **do**
    Calculate $f(x^i)$
end for
while termination criterion is not met and population has not yet converged **do**
    Empty the mating pool: $M = \{\}$
    Insert N individuals into M from P using a selection scheme
    Shuffle the position of all individuals in M
    $ind = 1$
    repeat
        if $random(0,1) \le p_c$ **then**
            Recombine $x^{ind} \in M$ and $x^{ind+1} \in M$ and place the resulting two offspring in P'
        else
            Copy $x^{ind} \in M$ and $x^{ind+1} \in M$ to P'
        end if
        $ind = ind + 2$
    until $ind > N$
    for $i = 1$ to N **do**
        for $j = 1$ to l **do**
            if $random(0,1) \le p_m$ **then**
                $mutate(x^i_j)$, where $x^i \in P'$
            end if
        end for
        Calculate $f(x^i)$, where $x^i \in P'$
    end for
    $P := P'$
end while

The following paragraphs briefly explain the basic elements of a GA. The selection process performed in population-based search approaches is equivalent to local search for single individuals as it distinguishes high-quality from low-quality solutions and selects promising solutions. Popular selection schemes are *proportionate selection* (Holland, 1975) and *tournament selection* (Goldberg et al, 1989). For proportionate selection, the expected number of copies a solution has in the next population is proportional to its fitness. The chance of a solution x^i being selected for recombination is calculated as

$$\frac{f(x^i)}{\sum_{j=1}^{N} f(x^j)}.$$

With increasing fitness, an individual is chosen more often for reproduction.

When using tournament selection, a tournament between s randomly chosen different individuals is held and the one with the highest fitness is added to the mating pool M. After N tournaments of size s the mating pool is filled. We have to distinguish between tournament selection with and without replacement. Tournament selection with replacement chooses for every tournament s individuals from the population P. Then, M is filled after N tournaments. Tournament without replacement performs s rounds. In each round we have N/s tournaments and we choose the solutions for a tournament from those who have not already taken part in a tournament in this round. After all solutions have performed a tournament in one round (after N/s tournaments), the round is over and all $x \in P$ are considered again for the next round. Therefore, to completely fill the mating pool, s rounds are necessary. For more information concerning different selection schemes, see Bäck et al (1997, C2) and Sastry and Goldberg (2001).

The mating pool M consists of all solutions which are chosen for recombination. When using tournament selection, there are no copies of the worst solution, and either an average of s copies (with replacement), or exactly s copies (without replacement) of the best solution.

Crossover operators imitate the principle of sexual reproduction and are applied to all $x \in M$. Usually, crossover produces two new offspring from two parents by exchanging substrings. Common crossover operators are one-point crossover and uniform crossover (Sect. 4.3, p. 117-120). The mutation operator is a local search operator and slightly changes the genotype of a solution $x \in P'$. It is important for local search, or if some alleles are lost during a GA run. Mutation can reanimate solution properties that have previously been lost. The probability of mutation p_m must be selected to be at a low level because otherwise mutation would randomly change too many alleles and new solutions would have nothing in common with their parents. Offspring would be generated almost randomly and genetic search would degrade to random search. For details on mutation operators, see Sect. 4.3.2, p. 115-117.

In GAs, intensification is mainly a result of the selection scheme. In a selection step, the average fitness of a population increases as only high-quality solutions are chosen for the mating pool M. Due to selection, the population converges after a number of generations (for details on the convergence speed see Thierens (1995), Goldberg (2002), or Suzuki (1995)). Continuing recombination-based search after the population has converged (hopefully to the global optimum) makes no sense as diversity is minimal.

The main source of diversification is the initial population. Therefore, in GAs, often large population sizes of a few hundreds or even thousands of solutions are used (Goldberg, 2002). The effect of recombination operators is twofold: On the one hand, recombination operators are able to create new solutions. On the other hand, usually recombination does not actively diversify the population but has an intensifying character. Crossover operators reduce diversity as the distances between offspring and parents are usually smaller than the distance between parents (see (4.1), p. 118). Therefore, the iterative application of crossover alone reduces the diversity of a population as either some solution properties can become extinct in the population (this effect is known as *genetic drift* (see p. 171) and especially relevant for binary solutions) or the decision variables converge to an average value (especially for continuous decision variables). The statistical properties of a population to which only recombination operators are applied, do not change (we have to consider that the aim of crossover is to re-combine different optimal solutions for subproblems) and, for example, the average fitness of a population remains constant (if we use a meaningful recombination operator and $N \rightarrow \infty$). However, diversity decreases as the solutions in the population become more similar to each other.

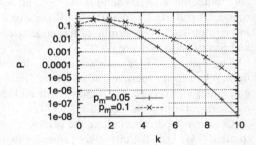

Fig. 5.4 Probability $P(k)$ of performing exactly k mutations versus k for $n = 20$ and various mutation rates p_m

In SGA, mutation has diversifying character. In contrast to many local search approaches where the neighborhood structure remains constant during search, the mutation operator used in SGA results in a varying neighborhood structure. Mutation does not generate only neighboring solutions with small distance but can reach all solutions in the search space in only one mutation step. When mutating one individual, mutation is iteratively applied to all l decision variables with probability p_m. Therefore, mutation can be modeled as a Bernoulli process and the probability of mutating exactly k decision variables can be calculated as $P(k) = \binom{n}{k}(p_m)^k(1 - p_m)^{n-k}$. Figure 5.4 plots $P(k)$ over k for $l = 20$ and $p_m = 0.05$ and $p_m = 0.1$. On average, $p_m l$ alleles are mutated and, for large values of p_m, the mutation operator can mutate all l decision variables and, thus, reach all possible points in the solution space. However, the probability of changing a large number of decision variables is low if p_m is low. The diversifying character of mutation increases with increasing p_m and for large p_m, SGA behaves like random search.

Further and more detailed information on functionality and application aspects of genetic algorithms can be found in standard textbooks like Goldberg (1989c), Mitchell (1996), Michalewicz (1996), Goldberg (2002), Michalewicz and Fogel (2004), Reeves and Rowe (2003), De Jong (2006), or Eiben and Smith (2010).

5.2.2 Estimation of Distribution Algorithms

Recombination operators should consider the linkage between different decision variables in a solution and transfer high-quality sub-solutions, which consist of several decision variables, as a whole from parents to offspring. However, the standard crossover operators like uniform or n-point crossover are position-dependent and thus sometimes of only limited use. n-point crossover can only work properly if the related decision variables are grouped together in the string. For example, for two-point crossover, sub-solutions for subproblems where the two decision variables x_0 and x_l are linked cannot be transferred as a whole from a parent to an offspring. In contrast, uniform crossover results in problems for large sub-problems as it disrupts such sub-solutions with high probability. For example, when using binary decision variables and assuming sub-problems of size k, uniform crossover disrupts sub-solutions with probability $1 - (1/2)^{k-1}$ $(k \geq 2)$.

To overcome the shortcomings of traditional crossover methods, two different approaches have been proposed in the literature: the first line of research modifies the genotypes and the corresponding representation so that linked decision variables are grouped together in the genotype. Then, traditional recombination operators like n-point crossover show good performance as they do not not disrupt sub-solutions. Examples are the *fast messy GA* (Goldberg et al, 1993a) or the *gene expression messy GA* (Bandyopadhyay et al, 1998).

The second line of research designs recombination operators that are able to detect the linkage between decision variables and construct new offspring position-independently considering the linkage. To avoid disruptions of sub-solutions, recombination is replaced by generating new solutions according to the probability distribution of promising solutions of the previous generation. Algorithms that follow this concept are called *estimation of distribution algorithms* (EDA) and have been introduced by Mühlenbein and Paaß (1996). EDAs are based on the same concepts as GAs but replace variation operators like crossover and mutation by sampling new solutions from a probabilistic model which takes into account the problem-specific interactions among the decision variables. Linkage between decision variables can be expressed explicitly through joint probability distributions associated with the variables of the individuals. For intensification and generation of high-quality solutions, EDAs use the same selection methods as GAs. Algorithm 8 outlines the basic functionality of EDAs.

Critical for the performance of EDAs is the proper estimation of the probability distribution because this process decomposes the problem into sub-problems by detecting a possible linkage between decision variables. This process can only work properly if the problem at hand is decomposable. Therefore, EDAs show good performance for problems that are decomposable and where the sub-problems are of small size. As for GAs, non-decomposable problems cannot be solved effectively by EDAs.

Different EDA variants can be classified with respect to the type of decision variables (binary versus continuous) and the probability distribution which is used to describe the statistical properties of a population. Simple variants for discrete prob-

---

Algorithm 8 Estimation of Distribution Algorithm

Create initial population P with N solutions x^i ($i \in \{1,\ldots,N\}$)
for $i = 1$ to N **do**
   Calculate $f(x^i)$
end for
while termination criterion is not met **do**
   Select M solutions, where $M < N$, from P using a selection scheme
   Estimate the (joint) probability distribution of all selected solutions
   Sample N solutions of the new population P' from the probability distribution
   Calculate $f(x^i)$, where $x^i \in P'$
   $P = P'$
end while

---

lem domains are *population-based incremental learning* (PBIL) (Baluja, 1994), *univariate marginal distribution algorithm* (UMDA) (Mühlenbein and Paaß, 1996), and *competent GA* (cGA) (Harik et al, 1998). These algorithms assume that all variables are independent (which is an unrealistic assumption for most real-world problems) and calculate the probability of an individual as the product of the probabilities of every decision variable. More advanced approaches assume bivariate dependencies between decision variables. Examples are *mutual information maximization for input clustering* (MIMIC) (de Bonet et al, 1997), *combining optimizers with mutual information trees* (COMIT) (Baluja and Davies, 1997), *probabilistic incremental program evolution* (PIPE) (Salustowicz and Schmidhuber, 1997), and *bivariate marginal distribution algorithm* (BMDA) (Pelikan and Mühlenbein, 1999). The most complex EDAs that assume multivariate dependencies between decision variables are the *factorized distribution algorithm* (FDA) (Mühlenbein and Mahnig, 1999), the *extended compact GA* (Harik, 1999), the *polytree approximation of distribution algorithm* (PADA) (Soto et al, 1999), *estimation of Bayesian networks algorithm* (Etxeberria and Larrañaga, 1999), and the *Bayesian optimization algorithm* (BOA) (Pelikan et al, 1999a). Although EDAs are still a young research field, EDAs using multivariate dependencies show promising results on binary problem domains and often outperform standard GAs (Larrañaga and Lozano, 2001; Pelikan, 2006). Their main advantage in comparison to standard crossover operators is the ability to learn the linkage between solution variables and the position-independent creation of new solutions.

The situation is different for continuous domains (Larrañaga et al, 1999; Bosman and Thierens, 2000; Gallagher and Frean, 2005). Here, the probability distributions used (for example Gaussian) are often not able to model the structure of the landscape in an appropriate way (Bosman, 2003) leading to a low performance of EDAs for continuous search spaces (Grahl et al, 2005).

For more detailed information on the functionality and application of EDAs we refer to Larrañaga and Lozano (2001) and Pelikan (2006).

5.2.3 Genetic Programming

Genetic programming (GP) (Smith, 1980; Cramer, 1985; Koza, 1992) is a variant
of GAs that evolves programs. Although most GP approaches use trees to represent
programs (Koza, 1992, 1994; Koza et al, 1999, 2005; Banzhaf et al, 1997; Langdon
and Poli, 2002), there are also a few approaches that encode programs using linear bitstrings (for example, *grammatical evolution* (Ryan, 1999; O'Neill and Ryan,
2003) or *Cartesian genetic programming* (Miller and Thomson, 2000)). The common feature of GP approaches is that the phenotypes are programs or variable-length
structures like electronic circuits or controllers.

In analogy to GAs, GP starts with a population of random candidate programs.
Each program is evaluated on a given task and its fitness value is assigned. Often,
the fitness of an individual is determined by measuring how well the found solution
(e.g. a computer program) performs a specific task. The basic functionality of GP
follows GA functionality. The main differences are the search space which consists
of tree structures of variable size and the corresponding search operators which have
to be tree-specific. Solutions (programs) are usually represented as *parse trees*. Parse
trees represent the syntactic structure of a string according to some formal grammar.
In a parse tree, each node is either a root node, a branch node, or a leaf node. Interior
nodes represent functions and leaf nodes represent variables, constants, or functions
with no arguments. Therefore, the nodes in a parse tree are either members of

- the *terminal set* T (leaf nodes) representing independent variables of the problem,
  zero-argument functions, random constants, or terminals with side-effects (for
  example move operations like "turn-left" or "move forward") or
- the *function set* F (interior nodes) representing functions (for example arithmetic
  or logical operations like "+", $\wedge$, or $\neg$), control structures (for example "if" or
  "while" clauses), or functions with side-effects (for example "write to file" or
  "read").

The definition of F and T is problem-specific and they should be designed such that

- each function is defined for each possible parameter. Parameters are either terminals or function returns.
- T and M must be chosen such that a parse tree can represent a solution for the
  problem. Solutions that cannot be constructed using the sets T and F are not
  elements of the search space and cannot be found by the search process.

In GP, the depth k and structure of a parse tree are variable. Figure 5.5 gives two
examples of parse trees.

The functionality of GP is analogous to GAs (see Algorithm 7, p. 148). Different
are the use of a direct representation (parse trees) and tree-specific initialization
and variation operators. During initialization, we generate random parse trees of
maximal depth k_{max}. There are three basic methods (Koza, 1992): *grow*, *full* and
ramped-half-and-half. The grow method starts with an empty tree and iteratively
assigns a node either to be a function or a terminal. If a node is a terminal, a random
terminal from the terminal set T is chosen. If the node is a function, we choose a

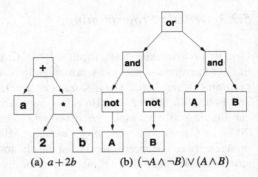

Fig. 5.5 Two examples of
parse trees

(a) $a + 2b$ (b) $(\neg A \wedge \neg B) \vee (A \wedge B)$

random function from F. Furthermore, a number of child nodes are added such that
their number equals the number of arguments necessary for the chosen function. The
procedure stops either at depth $k = k_{max}$ or when all leaf nodes are terminals. The
full method also starts with an empty tree but all nodes at depth $k \in \{1, \ldots, k_{max} - 1\}$
are functions. All nodes at depth $k = k_{max}$ become terminals. Therefore, all resulting
trees have depth k_{max}. For the ramped-half-and-half method, the population is evenly
divided into $(k_{max} - 1)$ parts. Half of each part is created by the grow method and
half by the full method. For both halves, the depth of the nodes in the ith part is
equal to i, where $i \in \{2, \ldots, k_{max}\}$. Thus, the diversity is high in the resulting initial
population.

As no standard search operators can be applied to parse trees, tree-specific
crossover and mutation operators are necessary. Like in SGAs, crossover is the
main search operator and mutation acts as background noise. Crossover exchanges
randomly selected sub-trees between two parse trees, whereas mutation usually replaces a randomly selected sub-tree by a randomly generated one. Like in GAs,
standard GP crossover (Koza, 1992) chooses two parent solutions and generates
two offspring by swapping sub-trees (see Fig. 5.6). Analogously, mutation chooses
a random sub-tree and replaces it with a randomly generated new one (see Fig. 5.7).
Other operators used in GP (Koza et al, 2005) are permutation, which changes the
order of function parameters, editing, which replaces sub-trees by semantically simpler expressions, and encapsulation, which encodes a sub-tree as a more complex
single node.

In recent years, GP has shown encouraging results finding programs or strategies
for problem-solving (Koza et al, 2005; Kleinau and Thonemann, 2004). However,
often the computational effort for finding high-quality solutions even for problems
of small sizes is extremely high. Currently, open problems in GP are the low locality
of the representation/operator combination (compare Chap. 7), the bias of the search
operators (compare the discussion in Sect. 6.2.3, p. 171) and *bloat*. Bloat describes
the problem that during a GP run, the average size of programs has been seen to
grow large, sometimes exponentially. Although there is a substantial amount of work
trying to fix problems with bloat (Nordin and Banzhaf, 1995; Langdon and Poli,
1997; Soule, 2002; Luke and Panait, 2006), there is no solution for this problem yet

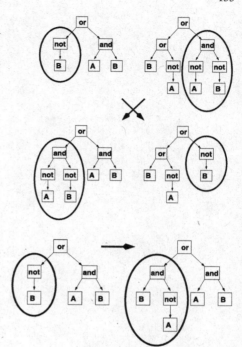

Fig. 5.6 Standard crossover
operator for GP

Fig. 5.7 Standard mutation
operator for GP

(Banzhaf and Langdon, 2002; Gelly et al, 2005; Luke and Panait, 2006; Whigham
and Dick, 2010).

Chapter 6
Design Principles

A fundamental assumption about the application of modern heuristics is that the vast majority of optimization problems that we can observe in the real world are

- neither deceptive nor difficult (Sect. 2.4.2.1) and
- have high locality (Sect. 2.4.2).

We assume that deceptive problems have no importance in the real world as usually optimal solutions are not isolated in the search space surrounded by only low-quality solutions. Furthermore, we assume that the metric of a search space is meaningful and, on average, the fitness differences between neighboring solutions are smaller than between randomly chosen solutions. Only because most real-world problems are neither difficult nor deceptive, guided search methods that use information about previously sampled solutions can outperform random search. This assumption is reasonable as the discussion about the NFL and search algorithms with subthreshold-seeking behavior (Sect. 3.4.4) shows that modern heuristics perform like random search on problems that are difficult or have a trivial topology (e.g. needle-in-a-haystack problems) and even worse on deceptive problems (e.g. traps).

Because we assume that the majority of problems in the real world are neither deceptive (we cannot solve such problems anyway) nor difficult (guided search methods cannot perform better than random search), we want to exclude deceptive as well as difficult problems from further consideration. We are not able to design reasonable search strategies for such types of problems. This leaves us with problems that are straightforward or, at least, straightforward in the area around the optimal solution (which also includes multi-modal problems). For such problems, guided search can be a reasonable strategy.

Since we assume that high locality is a general property of real-world problems, modern heuristics must ensure that their design does not destroy the high locality of a problem. If the high locality of a problem is destroyed, straightforward problems turn into difficult problems and cannot be solved better than by random search. Therefore, modern heuristics must ensure that the search operators used fit the met-

F. Rothlauf, *Design of Modern Heuristics*, Natural Computing Series, DOI 10.1007/978-3-540-72962-4_6, © Springer-Verlag Berlin Heidelberg 2011

ric on the search space and representations have high locality; this means phenotype distances must correspond to genotype distances.

The second topic of this chapter is how we can consider knowledge about problem-specific properties of a problem for the design of modern heuristics. For example, we have knowledge about the character and properties of high-quality (or low-quality) solutions. Such problem-specific knowledge can be exploited by introducing a *bias* into modern heuristics. The bias should consider this knowledge and, for example, concentrate search on solutions that are expected to be of high quality or avoid solutions expected to be of low quality. A bias can be considered in all design elements of modern heuristics namely the representation, the search operator, the fitness function, the initialization, and also the search strategy. However, modern heuristics should only be biased if we have obtained some particular knowledge about an optimization problem or problem instance. If we have no knowledge about properties of a problem, we should not bias modern heuristics as this will mislead the search heuristics.

Sect. 6.1 discusses how the design of modern heuristics can modify the locality of a problem. The locality of a problem is high if the distances between individuals correspond to fitness differences. To ensure guided search, the search operators must fit the problem metric. Local search operators must generate neighboring solutions and recombination operators must generate offspring where the distances between offspring and parents do not exceed the distances between parents. If the fit is poor, guided search is not possible any more and modern heuristics become random search due to a high diversification of search. Section 6.1.2 focuses on representations and recommends that the locality of representations should be high; this means neighboring genotypes correspond to neighboring phenotypes. Only then, the locality of a problem can be preserved and guided search becomes possible. Section 6.2 focuses on the possibility of biasing modern heuristics. We discuss how problem-specific construction heuristics can be used as genotype-phenotype mappings (Sect. 6.2.1) and how redundant representations affect heuristic search (Sect. 6.2.2). We distinguish between synonymously and non-synonymously redundant representations. Non-synonymously redundant representations lead to reduced performance of guided search methods for most real-world problems. Using synonymously redundant representations can increase performance if a priori knowledge is considered for their design. The section ends with comments on the bias of search operators (Sect. 6.2.3).

6.1 High Locality

The *locality* of a problem measures how well the distances $d(x,y)$ between any two solutions $x,y \in X$ correspond to the difference in their fitness values $|f(x) - f(y)|$ (Sect. 2.4.2). Locality is high if neighboring solutions have similar fitness values and fitness differences correlate positively with distances. In contrast, the locality of a problem is low if low distances do not correspond to low differences in the fitness values. Important for the locality of a problem is the metric defined on the search space. Usually, the problem metric is a result of the model developed for the problem.

The performance of guided search methods is high if the locality of a problem is relatively high, this means the structure of the fitness landscape leads search algorithms to high quality solutions (see Sects. 2.4.2 and 3.4.4). Local search methods show especially good performance if either high-quality or low-quality solutions are grouped together in the solution space (compare the discussion of the submedian-seeker in Sect. 3.4.4). Optimization problems with high locality can usually be solved well using modern heuristics as all modern heuristics have some kind of local search elements. Representative examples of optimization problems with high locality are unimodal problems like the sphere model (Sect. 5.1.5, p. 143) where high-quality solutions are grouped together in the search space.

This section discusses design guidelines for search operators and representations. Search operators must fit the metric of a search space, because otherwise modern heuristics show low performance as they behave like random search. A representation introduces an additional genotype-phenotype mapping. The locality of a representation describes how well the metric on the phenotype space fits to the metric on the genotype space. Low locality, which means there is a poor fit, randomizes the search and also leads to low performance of modern heuristics.

6.1.1 Search Operator

Modern heuristics rely on the concept of local search. Local search iteratively generates new solutions similar to existing ones. Local search is a reasonable and successful search approach for real-world problems as most real-world problems have high locality and are neither deceptive nor difficult. In addition, to avoid being trapped in local optima, modern heuristics use diversification steps. Diversification steps randomize search and allow modern heuristics to jump through the search space.

We have seen in the previous chapter that different types of modern heuristics use different concepts for controlling intensification and diversification. Local search intensifies the search as it allows incremental improvements of already found solutions. Diversification steps must be relatively rare as they usually lead to inferior solutions. When designing search operators, we must have in mind that modern heuristics use local search operators as well as recombination operators for *intensifying* the search (Sects. 5.1 and 5.2). Solutions that are generated should be similar to the existing ones. Therefore, we must ensure that search operators (local search operators as well as recombination operators) generate similar solutions and do not jump around in the search space. This can be done by ensuring that local search operators generate neighboring solutions and recombination operators generate solutions where the distances between parent and offspring are smaller or equal to the distances between parents (Sect. 4.3.3 and 4.1).

However, applying search operators to solutions defines a metric on the corresponding search space. With respect to the search operators, solutions are similar to each other if only a few local search steps suffice to transform one solution into another. Therefore, when designing search operators, it is important that the metric induced by the search operators fits the problem metric. If both metrics are similar (this means a local search operator creates neighboring solutions with respect to the problem metric), guided search will perform well as it can systematically explore promising areas of the search space.

Therefore, we should make sure that local search operators generate neighboring solutions. The fit between the problem metric and the metric induced by the search operators should be high. Then, most real-world problems, where neighboring solutions have on average similar fitness values, are easy to solve for modern heuristics.

The situation is reversed for deceptive problems where a negative correlation between distance and fitness difference exists (Fig. 2.3, p. 33). Then, search operators that fit the problem metric lead guided search methods away from the optimal solution and the performance of guided search methods is low. To solve such types of problems, search operators must neglect all distance information and jump around in the search space. Then, on average, guided search methods perform similarly to random search (see also the discussion in Sect. 2.4). However, as such problems are uncommon in the real world and optimization methods that show similar performance to random search are not effective, we do not recommend the use of search operators that do not fit the problem metric.

For real-world problems, the design or choice of proper search operators can be difficult if it is unclear what is a "natural" problem metric. We want to illustrate this issue for a scheduling problem. Given a number of tasks, we want to find an optimal schedule. There are different metrics that can be relevant for such a permutation problem. We have the choice between metrics based either on the relative or absolute ordering of the tasks (Sect. 4.3.5.5, p. 123). The choice of the "right" problem metric depends on the properties of the scheduling problem. For example, if we want to find an optimal class schedule, usually it is more natural to use a metric based on the absolute ordering of the tasks (classes). The relative ordering of the tasks is much less important as we have fixed time slots. The situation is reversed if we want to find an optimal schedule for a paint shop. For example, when painting different cars consecutively, color changes are time-consuming as paint tools have to be cleaned before a new color can be used. Therefore, the relative ordering of the tasks (paint jobs) is important as the tasks should be grouped together such that tasks that require the same color are painted consecutively and ordered such that the paint shop starts with the brightest colors and ends with the darkest ones.

This example makes clear that the most "natural" problem metric does not depend on the set of possible solutions but on the character of the underlying optimization problem and fitness function. The goal is to choose a metric such that the locality of the problem is high. The same is true for the design of operators. A high-quality local search operator should generate solutions with similar fitness. Then, problems become straightforward (Sect. 2.4, Fig. 2.3) and are easy to solve for modern heuristics.

6.1.2 Representation

Representations introduce an additional genotype-phenotype mapping and thus modify the fit between the metric on the genotype space (which is often induced by the search operators used), and the original problem metric on the phenotype space. High-quality representations ensure that the metric on the genotype space fits the original problem metric. The *locality of a representation* describes how well neighboring genotypes correspond to neighboring phenotypes (Rothlauf and Goldberg, 1999; Gottlieb and Raidl, 1999, 2000; Gottlieb et al, 2001; Rothlauf and Goldberg, 2000). In contrast to the locality of a problem, which measures the fit between fitness differences and phenotype distances, the locality of a representation measures the fit between phenotype distances and genotype distances.

The use of a representation can change the difficulty of problems (Sect. 4.2.2). For every deceptive or difficult problem $f(x) = f_p(f_g(x))$ there is a transformation T such that the problem $g(x) = f[T(x)]$ becomes straightforward (Liepins and Vose, 1990). Therefore, every misleading and difficult problem can be transformed into an easy problem by introducing a linear transformation T. In general, the difficulty of a problem can easily be modified by linear transformations T, which are equivalent to genotype-phenotype mappings. Liepins and Vose concluded that their results underscore the importance of selecting good representations, and that good representations are problem-specific.

The ability of representations to change the difficulty of a problem is closely related to their locality. The locality of a representation is high if all neighboring genotypes correspond to neighboring phenotypes. In contrast, it is low if neighboring genotypes do not correspond to neighboring phenotypes. Therefore, the locality d_m of a representation can be defined as (Rothlauf, 2006, Sect. 3.3)

$$d_m = \sum_{d_{x,y}^g = d_{min}^g} |d_{x,y}^p - d_{min}^p|, \tag{6.1}$$

where $d_{x,y}^p$ is the phenotype distance between the phenotypes x^p and y^p, $d_{x,y}^g$ is the genotypic distance between the corresponding genotypes, and d_{min}^p and d_{min}^g are the minimum distances between two (neighboring) phenotypes and genotypes, respectively. Without loss of generality, we assume that $d_{min}^g = d_{min}^p$. For $d_m = 0$, all genotypic neighbors correspond to phenotypic neighbors and the encoding has perfect (high) locality.

We want to emphasize that the locality of a representation depends on the representation f_g and the metrics that are defined on Φ_g and Φ_p. f_g alone only determines which phenotypes are represented by which genotypes and cannot be used for measuring similarity between solutions. To describe or measure the locality of a representation, a metric must be defined on Φ_g and Φ_p.

Figure 6.1 illustrates the difference between high and low-locality representations. We assume 12 different phenotypes (a-l) and measure distances between solutions using the Euclidean metric. Each phenotype (lower case symbol) corresponds to one genotype (upper case symbol). The representation f_g has perfect (high) lo-

cality if neighboring genotypes correspond to neighboring phenotypes. Then, local
search steps have the same effect in the phenotype and genotype search space.

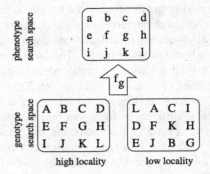

Fig. 6.1 High versus low-
locality representations

If we assume that f_g is a one-to-one mapping, every phenotype is represented by
exactly one genotype and there are $|\Phi_g|! = |\Phi_p|!$ different representations. Each of
these many different representations assigns the genotypes to the phenotypes in a
different way.

We want to ask how the locality of a representation influences the performance of
modern heuristics. Often, there is a "natural" problem metric describing which phenotypes are similar to each other. A representation introduces a new genotype metric
based on the genotypes and search operators used. This metric can be different from
the problem (phenotype) metric. Therefore, the character of search operators can be
different for genotypes versus phenotypes. If the locality of a representation is high,
then a search operator has the same effect on the phenotypes as on the genotypes. As
a result, the original problem difficulty remains unchanged by a representation
f_g. Easy (straightforward) problems remain easy and misleading problems remain
misleading. Figure 6.2 (left) illustrates the effect of local search operators for high-
locality representations. A local search step has the same effect on the phenotypes
as on the genotypes.

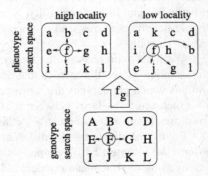

Fig. 6.2 The effect of local search operators for high
versus low-locality representations

For low-locality representations, the situation is different and the influence of a representation on the difficulty of a problem depends on its character. If a problem f_p is straightforward (we use the problem classification from Sect. 2.4.2, p. 32), a low-locality representation f_g randomizes the problem by destroying the correlation between distance and fitness and making the problem $f = f_p(f_g(x^g))$ more difficult. When using low-locality representations, a small change in a genotype does not correspond to a small change in the phenotype, but larger changes in the phenotype are possible (Fig. 6.2, right). Therefore, when using low-locality representations, straightforward problems become on average difficult as low-locality representations lead to a more uncorrelated fitness landscape and heuristics can no longer extract meaningful information about the structure of the problem. Guided search becomes more difficult as many genotypic search steps do not result in a similar solution but in a random one.

If the problem f_p is difficult, on average, a low-locality representation does not change the problem difficulty. Although local search turns into random search, the performance of modern heuristics stays constant as random search and local search show the same performance for difficult problems. Of course, representations exist that can make a particular problem easier and result in an overall straightforward problem; however, there are few of these and on average low-locality representations simply modify the problem but do not create a straightforward fitness landscape (Rothlauf, 2006, Sect. 3.3). On the other hand, there are also representations f_g that construct problems f which are misleading. But as for low-locality representations that transform a problem from difficult to straightforward, there are few of these representations.

Finally, we have to consider misleading problems. On average, the use of low-locality representations transforms such problems into difficult problems as the problems become more randomized. Then, local search is less misled by the fitness landscape and the performance of guided search methods improves. On average, low-locality representations reduce the "misleading character" of such problems and turn them into difficult problems (Rothlauf, 2006, Sect. 3.3).

Summarizing the results, low-locality representations have the same effect as using random search. Therefore, on average, straightforward problems become more difficult, and misleading problems become more easy to solve for guided search methods. As most real-world problems are straightforward, the use of low-locality representations makes these problems more difficult. Therefore, we strongly encourage users of modern heuristics to use high-locality representations for problems of practical relevance. Of course, low-locality representations make misleading problems easier for guided search; however, these are problems which we do not expect to meet in reality and we do not really want to solve.

For more information on the influence of the locality of representations on the performance of modern heuristics, we refer the interested reader to Rothlauf (2006, Sect. 3.3).

6.2 Biasing Modern Heuristics

This section discusses how to bias modern heuristics. If we have some knowledge about the properties of either high-quality or low-quality solutions, we can make use of this knowledge for the design of the key elements of modern heuristics. For representations, we can incorporate heuristics or introduce redundant encodings and assign a larger number of genotypes to high-quality phenotypes. Search operators can be designed in such a way that they distinguish between high-quality and low-quality solution features (building blocks) and prefer the high-quality ones. Analogously, we can bias the fitness function, the initial solution, and the search strategy.

A representation or search operator is *biased* if an application of a variation operator generates some solutions in the search space with higher probability (Caruana and Schaffer, 1988). We can bias representations by incorporating heuristics into the genotype-phenotype mapping. Furthermore, representations can be biased if the number of genotypes exceeds the number of phenotypes. Then, representations are called *redundant* (Gerrits and Hogeweg, 1991; Ronald et al, 1995; Shipman, 1999; Rothlauf and Goldberg, 2003). Redundant representations are biased if some phenotypes are represented by a larger number of genotypes. Analogously, search operators are biased if some solutions are generated with higher probability.

In analogy to representations and search operators, we can also bias the initialization method. An unbiased initialization method creates all solutions in the search space with equal probability and chooses the starting point of the search randomly. If we have some knowledge about the structure or properties of optimal solutions, we can bias the initial solutions and focus on more promising areas of the search space. Construction heuristics that apply some rules of thumb for the creation of an initial solution are usually biased as the created solution is expected to be better than an average solution. For more information and examples on biasing initial solutions, we want to refer to Sects. 4.5 and 8.5.

We can also bias the fitness function. Here, we must have in mind that the fitness function is biased per se as it allows us to distinguish high-quality from low-quality solutions. However, we can add an additional bias to the fitness function that favors solutions with advantageous properties. An example is the smoothing of fitness functions (Sect. 4.4). Here, we modify the fitness of solutions according to the fitness of neighboring solutions. An example of a search strategy with a systematic bias of the fitness function is GLS. GLS adds a penalty to all solutions that have some specific solution feature. Furthermore, we often bias the fitness function if infeasible solutions exist in the search space. Subject to the number or types of constraint violations, we add penalties to infeasible solutions. The penalties are added in such a way that they should lead guided search methods in the direction of feasible solutions (Gottlieb, 1999). For more information on bias of fitness functions, we refer to Sect. 4.4.

Finally, we also can bias a search strategy. Search strategies determine the intensification and diversification mechanisms of a modern heuristic. The ratio between intensification and diversification depends on the properties of the underlying problem. If we know that a problem is easy for pure local search (e.g. unimodal problems), we can bias a search strategy to use more intensification steps. For example, for unimodal problems, we can set the number of restarts for ILS to zero or start simulated annealing with temperature $T = 0$. If the problem is more difficult and a larger number of local optima exist, diversification becomes more important and we can bias modern heuristics to use more diversification steps. For example, we can increase the number of restarts for ILS or start SA with a higher temperature.

When biasing modern heuristics, we must make sure that we have a priori knowledge about the problem and the bias exploits this knowledge in an appropriate way. Introducing an inappropriate or wrong bias into modern heuristics would mislead search and result in low solution quality. Furthermore, we must make sure that a bias is not too strong. Using a bias can focus the search on specific areas of the search space and exclude solutions from consideration. If the bias is too strong, modern heuristics can easily fail. For example, reducing the number of restarts of ILS to zero and performing only local search is a promising strategy for unimodal problems. However, if only a few local optima exist, the bias is too strong as search can never escape any local optimum. In general, we must use bias carefully. A wrong or too strong bias can easily lead to failure of modern heuristics.

The following sections discuss bias of representations and search operators. Sect. 6.2.1 gives an overview of how problem-specific construction heuristics can be used as genotype-phenotype mappings. Then, heuristic search varies either the input (problem space search) or the parameters (heuristic space search) of the construction heuristic. Sect. 6.2.2 discusses redundant representations. Redundant representations with low locality (called *non-synonymously redundant* representations) randomize guided search and thus should not be used. Redundant representations with high locality (*synonymously redundant* representations) can be biased by overrepresenting solutions similar to optimal solutions. Finally, Sect. 6.2.3 discusses how the bias of search operators affects modern heuristics and give an example of the bias of standard search operators used in genetic programming.

6.2.1 Incorporating Construction Heuristics in Representations

We focus on combining problem-specific construction heuristics with genotype-phenotype mappings. The possibility to design problem-specific representations and to incorporate relevant knowledge about the problem into the genotype-phenotype mapping by using construction heuristics is a promising line of research and is continuously discussed in the operations research and evolutionary algorithm communities (Holland, 1975; Michalewicz and Janikow, 1989; Storer et al, 1995; Liepins and Vose, 1990; Michalewicz and Schoenauer, 1996; Surry, 1998).

Genotype-phenotype mappings map genotypes to phenotypes and can incorporate problem-specific construction heuristics. An early example of a problem-specific representation is the *ordinal representation* of Grefenstette et al (1985) who studied the performance of genetic algorithms for the TSP. The ordinal representation encodes a tour (permutation of n integers) by a genotype x^g of length n, where

$x_i^g \in \{1,\ldots,n-i\}$ and $i \in \{0,\ldots,n-1\}$. For constructing a phenotype, a predefined permutation x^s of n integers representing the n different cities is used. x^s can
be problem-specific and, for example, consider edge weights. A phenotype (tour) is
constructed from x^g by subsequently adding (starting with $i = 0$) the x_i^gth element
of x^s to the phenotype (which initially contains no elements) and removing the x_i^gth
element of x^s. Problem-specific knowledge can be considered by choosing an appropriate x^s as genotypes define perturbations of x^s and using small integers for the
x_i^g results in a bias of the resulting phenotypes towards x^s. For example, for $x_i^g = 1$
($i \in \{0,\ldots,n-1\}$), the resulting phenotype is x^s.

Other early examples of problem-specific representations are Bagchi et al (1991)
who incorporated a problem-specific schedule builder into representations for job
shop scheduling problems, Jones and Beltramo (1991) who studied representations
that use a greedy adding heuristic for partitioning problems, and the more general *Adaptive Representation Genetic Optimization Technique* (*ARGOT*) strategy
(Shaefer, 1987; Shaefer and Smith, 1990). ARGOT dynamically changes either the
structure of the genotypes or the genotype-phenotype mapping according to the
progress made during search. Examples of problem-specific tree representations are
presented in Sect. 8.1.2.

In parallel to, and independently from representations, Storer et al (1992) proposed *problem space search* (PSS) and *heuristic space search* (HSS) as approaches
that also combine problem-specific heuristics with problem-independent modern
heuristics. PSS and HSS apply in each search iteration of a modern heuristic a
problem-specific base heuristic H that exploits known properties of the problem.
The base heuristic H should be fast and creates a phenotype from a genotype. Results presented for different applications show that this approach can lead to improved performance of modern heuristics (Storer et al, 1995; Naphade et al, 1997;
Ernst et al, 1999). For PSS, H is applied to perturbed versions of the genotype. The
perturbations of the genotypes are usually small and based on a definition of neighborhood in the genotype space. For HSS, in each search step of the modern heuristic,
the genotypes remain unchanged, but (slightly) different variants of the base heuristic H are used. For scheduling problems, linear combinations of priority dispatching
rules with different weights, or the application of different base heuristics to different parts of the genotype, have been proposed (Naphade et al, 1997).

PSS and HSS use the same underlying concepts as problem-specific representations. The base heuristic H is equivalent to a (usually problem-specific) genotype-
phenotype mapping and assigns phenotypes to genotypes. PSS performs heuristic
search by modifying (perturbing) the genotypes, which is equivalent to the concept of representations originally proposed by Holland (1975) (for an early example
see Grefenstette et al (1985)). HSS perturbs the base heuristic (genotype-phenotype
mapping) which is similar to the concept of adaptive representations (for early examples see Shaefer (1987) or Schraudolph and Belew (1992)).

6.2.2 Redundant Representation

We assume a combinatorial optimization problem with a finite number of pheno-types. If the size of the genotype search space is equal to the size of the phenotype search space ($|\Phi_g| = |\Phi_p|$) and the representation maps all genotypes to all pheno-types (bijection), a representation cannot be biased. All solutions are represented with the same probability and a bias can only be a result of the search operator used.

The situation is different for representations where the number of genotypes exceeds the number of phenotypes. We still assume that all phenotypes are en-coded by at least one genotype. Such representations are usually called *redundant representations* (e.g. in Cohoon et al (1988), Ronald et al (1995), Julstrom (1999), or Rothlauf and Goldberg (2003)). Radcliffe and Surry (1994) introduced a differ-ent notion of redundancy and distinguished between *degenerated representations*, where more than one genotype encodes one phenotype, and redundant representa-tions where parts of the genotypes are not used for the construction of a phenotype. However, this distinction has not generally been accepted in the modern heuristic community. Therefore, we follow the majority of the literature and define encod-ings to be redundant if the number of genotypes exceeds the number of phenotypes (which is equivalent to the notion of degeneracy of Radcliffe and Surry (1994)).

Rothlauf and Goldberg (2003) distinguished between different types of redun-dant representations based on the similarity of the genotypes that are assigned to the same phenotype. A representation is defined to be *synonymously redundant* if the genotypes that are assigned to the same phenotype are similar to each other. Con-sequently, representations are *non-synonymously redundant* if the genotypes that are assigned to the same phenotype are not similar to each other. Therefore, the synonymity of a representation depends on the genotype and phenotype metric. Fig-ure 6.3 illustrates the differences between synonymous and non-synonymous redun-dant encodings. Distances between solutions are measured using a Euclidean met-ric. The symbols indicate different genotypes and their corresponding phenotypes. When using synonymously redundant representations (left), genotypes that repre-sent the same phenotype are similar to each other. When using non-synonymously redundant representations (right), genotypes that represent the same phenotype are not similar to each other but distributed over the whole search space.

Formally, a redundant representation f_g assigns a phenotype x^p to a set of dif-ferent genotypes $x^g \in \Phi_g^{x^p}$, where $\forall x^g \in \Phi_g^{x^p} : f_g(x^g) = x^p$. All genotypes x^g in the genotype set $\Phi_g^{x^p}$ represent the same phenotype x^p. A representation is synony-mously redundant if the genotype distances between all $x^g \in \Phi_g^{x^p}$ are small for all different x^p. Therefore, if for all phenotypes the sum over the distances between all genotypes that correspond to the same phenotype

$$\sum_{x^p} \frac{1}{2} \left(\sum_{x^g \in \Phi_g^{x^p}} \sum_{y^g \in \Phi_g^{x^p}} d(x^g, y^g) \right), \tag{6.2}$$

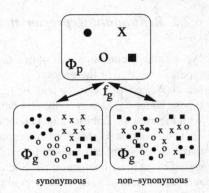

Fig. 6.3 Synonymous versus
non-synonymous redundancy

where $x^g \neq y^g$, is reasonably small, a representation is called synonymously redundant. $d(x^g, y^g)$ depends on the metric used and measures the distance between two genotypes $x^g \in \Phi_g^{x^p}$ and $y^g \in \Phi_g^{x^p}$ which both represent the same phenotype x^p.

6.2.2.1 Non-Synonymously Redundant Representations

The synonymity of a representation can have a large influence on the performance of modern heuristics. When using non-synonymously redundant representations, a local search operator can result in solutions that are phenotypically completely different from their parents. For recombination operators, the distances between offspring and parents are not necessarily smaller than the distances between parents.

Local search methods like the submedian-seeker outperform random search if solutions with similar fitness are grouped together in the search space and are not scattered over the whole search space (Radcliffe, 1991a; Manderick et al, 1991; Horn, 1995; Deb et al, 1997; Christensen and Oppacher, 2001; Whitley, 2002) (Sect. 3.4.4). Furthermore, problems are easy for guided search methods if distances between solutions are related to corresponding fitness differences (Sect. 2.4.2). However, non-synonymous redundant representations destroy existing correlations between solutions and their corresponding fitness values. Thus, search heuristics cannot use any information learned during the search for determining future search steps. As a result, it makes no sense for guided search approaches to search around already found high-quality genotypes and guided search algorithms become random search. A local search step does not result in a solution with similar properties but in a random solution. Analogously, recombination is not able to create new solutions with similar properties to their parents, but creates new, random solutions. Therefore, non-synonymously redundant representations have the same effect on modern heuristics as low-locality representations (Sect. 6.1.2).

The use of non-synonymously redundant representations allows us to reach many different phenotypes in a single local search step (Fig. 6.4, right). However, increasing the *connectivity* between phenotypes results in random search and decreases the efficiency of modern heuristics. As for low-locality representations, a genotype search step does not result in a similar phenotype but creates a random solution. Therefore, guided search is no longer possible but becomes random search. As a result, we get reduced performance of modern heuristics on straightforward problems when using non-synonymously redundant representations.

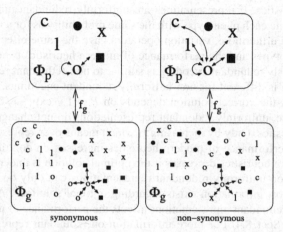

Fig. 6.4 The effects of local search steps for synonymously versus non-synonymously redundant representations. The arrows indicate search steps

On average, non-synonymously redundant representations transform straightforward as well as misleading problems into difficult problems where the fitness differences between two solutions are not correlated to their distances. Easy problems become more difficult and deceptive problems more easy. Therefore, we do not recommend using non-synonymously redundant encodings. A more detailed discussion of non-synonymously redundant representations can be found in Rothlauf (2006, Sect. 3.1) and Choi and Moon (2008).

6.2.2.2 Bias of Synonymously Redundant Representations

The use of synonymously redundant representations allows local search to generate neighboring solutions. Small variations of genotypes cannot result in large phenotypic changes but result either in the same, or a similar, phenotype (Fig. 6.4, left).

To describe relevant properties of synonymously redundant representations, we can use the *order of redundancy*, k_r, which is defined as $k_r = \log(|\Phi_g|)/\log(|\Phi_p|)$ (Rothlauf, 2006, Sect. 3.1). k_r measures the amount of redundant information in the encoding. Furthermore, we are especially interested in biases of synonymously redundant representations. r measures a bias and denotes the number of genotypes that represent the optimal solution. When using non-redundant representations, every phenotype is assigned to exactly one genotype ($r = 1$). In general, $1 \le r \le |\Phi_g| - |\Phi_p| + 1$.

Synonymously redundant representations are unbiased (*uniformly redundant*) if each phenotype is on average encoded by the same number of genotypes. In contrast, encodings are biased (*non-uniformly redundant*) if some phenotypes are represented by a different number of genotypes. Rothlauf and Goldberg (2003) studied how the bias of synonymously redundant representations influence the performance of modern heuristics. If representations are uniformly redundant, unbiased search operators generate each phenotype with the same probability as for a non-redundant representation. Furthermore, variation operators have the same effect on the genotypes and phenotypes and the performance of modern heuristics using a uniformly and synonymously redundant encoding is similar to non-redundant representations.

The situation is different for non-uniformly redundant encodings. The probability P of finding the correct solution depends on $P \propto 1 - \exp(-r/2^{k_r})$ (Rothlauf, 2006). Therefore, uniformly redundant representations do not change the behavior of modern heuristics. Only by increasing r, which means over-representing optimal solutions, the performance of modern heuristics increases. In contrast, the performance of modern heuristics decreases if the optimal solution is underrepresented.

Therefore, non-uniformly redundant representations can only be used advantageously if a-priori information exists regarding optimal solutions. An example of how the use of redundant representations affects the performance of modern heuristics is given in Sect. 8.4. For more information on redundant representations, we refer to Rothlauf and Goldberg (2003).

6.2.3 Search Operator

Search operators are applied either to genotypes or phenotypes and subsequently create new solutions. Search operators can be either *biased* or *unbiased*. In the unbiased case, each solution in the search space has the same probability of being created. If some phenotypes have higher probabilities to be created by applying a search operator to a randomly chosen solution, we call this a *bias* towards those phenotypes (Rothlauf, 2006; Raidl and Gottlieb, 2005).

Using biased search operators can be helpful if some knowledge about the structure of high-quality solutions exists and the search operators are biased such that high-quality solutions are preferred. Then, the average fitness of solutions that are generated by the biased search operator is higher than randomly generated solutions or search operators without a bias.

Problems can occur if a bias induced by a search operator is either too strong or towards solutions that have a large distance to an optimal solution. If the bias is too strong, the diversity of a population is reduced and the individuals in a population quickly converge towards those solutions to which the search operators are biased. Then, after a few search steps, no heuristic search is possible any more (for an example, see the heuristic crossover variants in Fig. 8.4, Sect. 8.3.2). Furthermore, the performance of modern heuristics decreases if a bias exists towards solutions that have a large distance to optimal solutions. The biased search operators push a population of solutions in the "wrong direction" and it is difficult for modern heuristics to find optimal solutions. Therefore, biased search operators should be used with care.

Standard search operators for binary, integer, or continuous genotypes (see Sect. 4.3.5) are unbiased. The situation is slightly different for standard recombination operators. Recombination operators never introduce new solution features but only recombine existing properties. Thus, once some solution features are lost in a population of solutions they can never be regained later by using recombination operators alone. Given a randomly generated and unbiased initial population of solutions, the iterative application of recombination operators can result in a random fixation of some decision variables, which reduces the diversity of solutions that exist in a population. This effect is known as *genetic drift*. The existence of genetic drift is widely known and has been addressed in the field of population genetics (Kimura, 1962, 1964; Gale, 1990; Nagylaki, 1992; Hartl and Clark, 1997), and also in the field of evolutionary algorithms (Goldberg and Segrest, 1987; Asoh and Mühlenbein, 1994; Thierens et al, 1998; Lobo et al, 2000).

When using more sophisticated phenotypes and direct search operators, identifying bias of search operators may be difficult. For example, in standard genetic programming approaches, programs and syntactical expressions are encoded as trees of variable size. Daida et al (2001) showed that the two standard search operators in genetic programming (sub-tree swapping crossover and sub-tree mutation) are biased as they do not effectively search all tree shapes. In particular, very full or very narrow tree solutions are extraordinarily difficult to find, even when the fitness function provides good guidance to the optimum solutions. Therefore, genetic programming approaches will perform poorly on problems where optimal solutions require full or narrow trees. Furthermore, since the search operators do not find solutions that are at both ends of this "fullness" spectrum, problems may arise if we use those search operators to solve problems whose solutions are restricted to a particular shape, of whatever degree of fullness. Hoai et al (2006) studied approaches to overcome these problems and introduced a new tree-based representation and local insertion and deletion search operators with a lower bias.

In general, the bias of search operators can be measured by comparing the properties of randomly generated solutions with solutions that are created by subsequent applications of search operators. Examples of how to analyze the bias of search operators are given in Sect. 8.3.

Part III
Case Studies

Chapter 7
High Locality Representations for Automated Programming

The previous chapter illustrated that high locality of representations and search operators is a prerequisite for efficient and effective modern heuristics. This chapter presents a case study on the locality of the genotype-phenotype mapping of grammatical evolution in comparison to standard genetic programming. *Grammatical evolution* (GE) is a variant of GP (Sect. 5.2.3, p. 153) that can evolve complete programs in an arbitrary language using a variable-length binary string (O'Neill and Ryan, 2001). In GE, phenotypes (programs) are created from binary genotypes by using a complex representation (genotype-phenotype mapping). The representation selects production rules in a Backus-Naur form grammar and thereby creates a phenotype. GE approaches have been applied to several test problems and real-world applications and good performance has been reported (O'Neill and Ryan, 2001, 2003; Ryan and O'Neill, 1998).

We study the locality of the genotype-phenotype mapping used in GE. In contrast to standard GP, which applies search operators directly to phenotypes (Sect. 5.2.3), GE uses an additional mapping and applies search operators to binary genotypes. Therefore, there is a large semantic gap between genotypes (binary strings) and phenotypes (programs or expressions). The case study shows that the mapping used in GE has low locality leading to low performance of standard mutation operators. The study at hand is an example of how basic design principles of modern heuristics can be applied to explain performance differences between different GP approaches and demonstrates current challenges in the design of GE.

We start with relevant properties of GE. Then, Sect. 7.2 presents two standard benchmark problems for GP. In Sect. 7.3, we study the locality of the genotype-phenotype mapping used in GE. Finally, Sect. 7.4 presents results on how the locality of the representation affects the performance of local search.

F. Rothlauf, *Design of Modern Heuristics*, Natural Computing Series,
DOI 10.1007/978-3-540-72962-4_7, © Springer-Verlag Berlin Heidelberg 2011

7.1 Grammatical Evolution

Grammatical evolution is a form of linear GP that differs from traditional GP (Sect. 5.2.3) in three ways:

- it employs linear genomes,
- it uses a grammar in Backus-Naur form (BNF) to define phenotype structures,
- and it performs an ontogenetic mapping from the genotype to the phenotype.

Altogether, these three characteristics enable GE to evolve complete computer programs in an arbitrary language using a variable-length binary genotype string.

Functionality

GE is a search method that can evolve computer programs defined in BNF. In contrast to standard GP (Koza, 1992), the genotypes are not parse trees but bitstrings of variable length. A genotype consists of groups of eight bits (called *codons*) that select production rules from a BNF grammar. The construction of the phenotype from the genotype is presented in Sect. 7.1.

The functionality of GE follows standard GA approaches using binary genotypes (Algorithm 7, p. 148). As standard binary strings are used as genotypes, no specific crossover or mutation operators are necessary. Therefore, we have an indirect representation and standard crossover operators like one-point or uniform crossover and standard mutation operators like bit-flipping mutation can be used (Sect. 4.3.5). A common metric for measuring distances between binary genotypes is Hamming distance (2.7). Therefore, the application of bit-flipping mutation creates a new solution with genotype distance $d^g = 1$. For selection, standard operators like tournament selection or roulette-wheel selection can be used (Sect. 5.2.1). Some GE implementations use steady state replacement mechanisms and duplication operators that duplicate a random number of codons and insert these after the last codon position. As usual, selection decisions are performed based on the fitness of phenotypes.

GE has been successfully applied to a number of diverse problem domains such as symbolic regression (O'Neill and Ryan, 2001, 2003), trigonometric identities (Ryan and O'Neill, 1998), symbolic integration (O'Neill and Ryan, 2003), the Santa Fe trail (O'Neill and Ryan, 2001), and others. The results indicate that GE can be applied to a wide range of problems and validate the ability of GE to generate multi-line functions in any language following BNF notation.

Backus-Naur Form

In GE, the *Backus-Naur form grammar* is used to define the grammar of a language as *production rules*. Based on the information stored in the genotypes, BNF-production rules are selected and form the phenotype. BNF distinguishes between terminals, which are equivalent to leaf nodes in trees, and non-terminals, which can be interpreted as expandable interior nodes (Sect. 5.2.3). A *grammar* in BNF is defined by the set $\{N,T,P,S\}$, where N is the set of non-terminals, T is the set of terminals, P is a set of production rules that maps N to T, and $S \in N$ is a start symbol.

In order to apply GE to a problem, it is necessary to define a BNF grammar for the problem. It must be defined in such a way that an optimal solution can be created from the elements defined by the grammar.

Genotype-phenotype Mapping

In GE, a phenotype is created from binary genotypes in two steps. In the first step, integer values are calculated from codons of eight bits. Therefore, from a binary genotype $x^{g,bin}$ of length $8l$, we get an integer genotype $x^{g,int}$ of length l, where each integer $x_i^{g,int} \in \{0,\ldots,255\}$, for $i \in \{0,\ldots,l-1\}$. Beginning with the start symbol $S \in N$, the integer value $x_i^{g,int}$ is used to select production rules from the BNF grammar. n_P denotes the number of production rules in P. To select a rule, we calculate the number of the used rule as $x_i^{g,int} \mod n_P$ (mod denotes the modulo operation). In this manner, the mapping process traverses the genotype string beginning from the left hand side ($x_0^{g,int}$) until one of the following situations arises:

- The mapping is complete. All non-terminals are transformed into terminals and a complete phenotype x^p is generated.
- The end of the genome is reached ($i = l - 1$) but the mapping process is not yet finished. The individual is wrapped, which means the alleles are reused, and the reading of codons continues. As the genotype is iteratively used with different meaning, wrapping can have a negative effect on locality. However, without wrapping, a larger number of individuals would be incomplete and infeasible.
- An upper threshold on the number of wrapping events is reached and the mapping is not yet complete. The mapping process is halted and the individual is assigned the lowest possible fitness value.

The mapping is deterministic, as the same genotype always results in the same phenotype. However, the interpretation of $x_i^{g,int}$ can be different if the genotype is wrapped and a different type of rule is selected. Figure 7.1 illustrates the functionality of the mapping. One after another, eight bits of $x^{g,bin}$ are interpreted as a binary string resulting in an integer value between 0 and 255. Then, we iteratively select the ($x^{g,int} \mod n_p$)th production rule. For the example, we assume $n_p = 4$.

For a more detailed description of the mapping process including illustrative examples, we refer to O'Neill and Ryan (2001) or O'Neill and Ryan (2003).

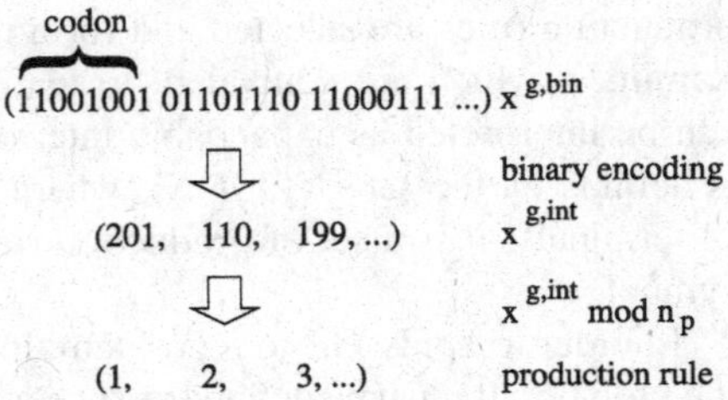

codon

$$\overbrace{(11001001\ 01101110\ 11000111 \ldots)}\ x^{g,bin}$$

⇩   binary encoding

$$(201,\quad 110,\quad 199, \ldots)\qquad x^{g,int}$$

Fig. 7.1 Functionality of
genotype-phenotype mapping
in GE

⇩ $x^{g,int}\ mod\ n_p$

$$(1,\quad 2,\quad 3, \ldots)\qquad \text{production rule}$$

7.2 Test Problems

We study the locality and performance of GE for the Santa Fe ant trail and symbolic
regression problem. Both problems are commonly used test problems for GP and
GE.

Santa Fe Ant Trail

In the Santa Fe Ant trail problem, 89 pieces of food are located on a discontinuous
trail which is embedded in a 32×32 toroidal grid. The goal is to determine rules that
guide the movements of an artificial ant and allow the ant to collect the maximum
number of pieces of food in t_{max} search steps. In each search step, exactly one action
can be performed. The ant can turn left (`left()`), turn right (`right()`), move
one square forward (`move()`), or look ahead one square in the direction it is facing
(`food_ahead()`). Figure 7.2(a) shows the BNF grammar for the Santa Fe ant trail
problem.

Symbolic Regression

In this example (Koza, 1992), a mathematical expression in symbolic form must be
found that approximates a given set of 20 data points (x_i, y_i). The function to be
approximated is

$$f(x) = x^4 + x^3 + x^2 + x,\tag{7.1}$$

where $x \in [-1; 1]$. A corresponding BNF grammar is shown in Fig. 7.2(b).

7.3 Locality of Grammatical Evolution

To measure the locality of a representation, we need a metric on the genotype and
phenotype space. For binary genotypes, often the Hamming distance (2.8) is used. It
measures the number of different bits in two genotypes x^g and y^g and is calculated as

```
N= {code,line,expr,if-stat,op}        N= {expr, op, pre_op}
T= {left(), right(), move(),          T= {sin,cos,exp,log,+,-,/,*,x,1,(,)}
   food_ahead(), else, if, {,         S= <expr>
   }, (,), ;}                         Production rules P:
 S= code                              <expr>    ::= <expr><op><expr>
Production rules P:                             |  (<expr><op><expr>)
<code>      ::= <line>                           |  <pre-op>(<expr>)
            |   <code><line>                     |  <var>
<line>      ::= <expr>                 <op>      ::= +
<expr>      ::= <if-stat>                        |  -
            |   <op>                             |  /
<if-stat> ::= if(food_ahead())                  |  *
               {<expr>} else          <pre-op> ::= sin
               {<expr>}                         |  cos
<op>        ::= left();                         |  exp
            |   right();                        |  log
            |   move();                <var>     ::= x
                                                |  1
        (a)  Santa Fe Ant trail                      (b)  symbolic regression
```

Fig. 7.2 BNF grammars for test problems

$d^g_{x^g,y^g} = \sum_i |x^g_i - y^g_i|$. A mutation (bit-flipping) of a solution x results in a neighboring solution y with distance $d^g_{x,y} = 1$.

Tree Edit Distance

It is more difficult to define appropriate metrics for phenotypes that are programs or expressions. In GE and GP, phenotypes are usually described as trees. Therefore, edit distances can be used for measuring differences/similarities between phenotypes. In general, the edit distance between two trees (phenotypes) is defined as the minimum cost sequence of elemental edit operations that transform one tree into the other. There are three elemental operations:

- deletion: A node is removed from the tree. The children of this node become children of its parent.
- insertion: A single node is added.
- replacement: The label of a node is changed.

To every operation a cost is assigned (usually the same for the different operations). Selkow (1977) presented an algorithm to calculate an edit distance where the operations insertion and deletion may only be applied to leaf nodes. Tai (1979) introduced an unrestricted edit distance and Zhang and Shasha (1989) developed a dynamic programming algorithm to compute tree edit distances.

In the context of GP, tree edit distances have been used as a measurement for the similarity of trees (Keller and Banzhaf, 1996; O'Reilly, 1997; Brameier and

Banzhaf, 2002). Igel (1998) and Igel and Chellapilla (1999) used tree edit distances for analyzing the locality of GP approaches.

Results

To investigate the locality of the genotype-phenotype mapping used in GE, we created 1,000 random genotypes. For the genotypes, we used standard parameter settings. The length of a binary genotype is 160 bits, the codon size is 8, the wrapping operator is used, the upper bound for wrapping events is 10, and the maximum number of elements in the phenotype is 1,000. For each individual x, we created all 160 neighbors y, where $d_{x,y}^g = 1$. The neighbors differ in exactly one bit from the original solution. The locality of the genotype-phenotype mapping can be determined by measuring the distance $d_{x,y}^p$ between the phenotypes that correspond to the neighboring genotypes x and y. The phenotype distance $d_{x,y}^p$ is measured as the edit distance between x^p and y^p.

For the GE genotype-phenotype mapping, we use the version 1.01 written by Michael O'Neill (O'Neill, 2006). The GE representation also contains the BNF Parser Grammar, version 0.63 implemented by Miguel Nicolau. For calculating the tree edit distance, we used a dynamic programming approach implemented by Zhang and Shasha (1989).

As the representation used in GE is redundant, some changes of the genotypes may not affect the corresponding phenotypes. We performed experiments for the Santa Fe ant trail problem and the symbolic regression problem and found that either 82% (Santa Fe) or 94% of all genotype neighbors are phenotypically identical ($d_{x,y}^p = 0$). Therefore, in about 90% of cases a mutation of a genotype (resulting in a neighboring genotype) does not change the corresponding phenotype.

Important for the locality of GE are the remaining neighbors that result in different phenotypes. The locality is high if the corresponding phenotypes are similar to each other. Figure 7.3 shows the frequency and cumulative frequency over the distance $d_{x,y}^p$ for the two different test problems. We only consider the case where $d_{x,y}^p > 0$. The results show that for the Santa Fe ant trail problem, many genotype neighbors are also phenotype neighbors (about 78%). However, there is also a significant number of genotype neighbors where the corresponding phenotypes are completely different. For example, more than 8% of all genotype neighbors have a tree edit distance $d_{x,y}^p \geq 5$. The situation is worse for symbolic regression. Only about 45% of all genotype neighbors correspond to phenotype neighbors and about 14% of all genotype neighbors correspond to phenotypes where $d_{x,y}^p \geq 5$.

We see that the locality of the genotype-phenotype mapping used in GE is not perfect. For the two test problems, a substantial percentage of neighboring genotypes do not correspond to neighboring phenotypes. Therefore, we expect some problems for the performance of local search approaches.

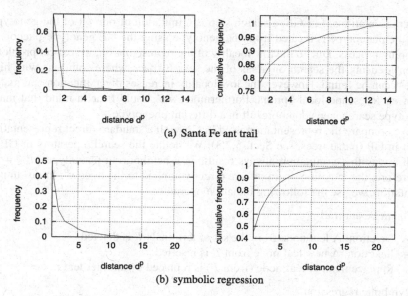

(a) Santa Fe ant trail

(b) symbolic regression

Fig. 7.3 Distribution of tree edit distances $d_{x,y}^p$ for neighboring genotypes x and y, where $d_{x,y}^g = 1$. We show the frequency (left) and cumulative frequency (right) over $d_{x,y}^p$ for the Santa Fe ant trail problem and the symbolic regression problem

7.4 Influence of Locality on GE Performance

The previous results indicate some problems of GE with low locality. Therefore, we investigate how strongly the low locality of the genotype-phenotype mapping influences the performance of GE. We focus the study on mutation only. However, the results for mutation are usually also relevant for crossover operators (Gottlieb et al, 2001; Gottlieb and Raidl, 1999; Rothlauf, 2006).

Experimental Setting

For the experiments, we want to make sure that we only examine the impact of locality on GE performance and that no other factors blur the results. Therefore, we implement a simple local search using only mutation as a search operator. The search strategy starts with a randomly created genotype and iteratively applies one bit-flipping mutation to the genotype. If the offspring has a higher fitness than the parent, it replaces it. Otherwise, the parent remains the actual solution.

We perform experiments for both test problems and compare an encoding with high locality with the representation used in GE. In the runs, we randomly generate a GE-encoded initial solution and use this solution as the initial solution for both types of representations. For GE, a search step is a mutation of one bit of the genotype, and the phenotype is created from the genotype using the GE genotype-phenotype mapping process. Due to the low locality of the representation, we expect problems when focusing the search on areas of the search space where solutions with high fitness can be found. However, the low locality increases diversification and makes it easier to escape local optima. Furthermore, we should bear in mind that many genotype search steps do not result in a different phenotype.

We compare the representation used in GE with a standard direct representation used in GP (parse trees, see Sect. 5.2.3). We define the search operators of GP in such a way that a mutation always results in a neighboring phenotype ($d^p_{x,y} = 1$). Therefore, the mutation operators are directly applied to the trees x^p. In our implementation, we use the following mutation operators:

- Santa Fe ant trail

  - Deletion: A leaf node from the set of terminals T is deleted.
  - Insertion: A new leaf node from T is inserted.
  - Replacement: A leaf node (from T) is replaced by another leaf node.

- symbolic regression

  - Insertion: sin, cos, exp, or log are inserted at a leaf that contains x.
  - Replacement: +, -, *, and / are replaced by each other; sin, cos, exp, and log are replaced by each other.

In each search step, the type of mutation operator used is chosen randomly. A mutation step always results in a neighboring phenotype and we do not need an additional genotype-phenotype mapping like in GE as we apply the search operators directly to the phenotypes.

Comparing these two different approaches, in GE, a mutation of a genotype results in most cases in the same phenotype, sometimes in a neighboring phenotype, but also sometimes in phenotypes that are completely different (see Fig. 7.3). In contrast, the standard GP representation (parse trees) is a high-locality "representation" as mutation always results in a neighboring phenotype. Therefore, the search can focus on promising areas of the search space but never escape local optima.

Performance Results

For the GE approach, we use the same parameter setting as described in Sect. 7.3. For both problems, we perform 1,000 runs of a local search strategy using randomly created initial solutions. Each run is stopped after 1,000 search steps. Figure 7.4 compares the performance for the Santa Fe ant trail (Fig. 7.4(a)) and the symbolic regression problem (Fig. 7.4(b)) over the number of search steps. Figure 7.4(a) shows the mean fitness of the found solution and Fig. 7.4(b) shows the mean error $1/20 \sum_{i=0}^{19} |f_j(x_i) - f(x_i)|$), where f is defined in (7.1) and f_j ($j \in \{0, \ldots, 1000\}$)

denotes the function found by the search in search step j. The results are averaged over all 1,000 runs.

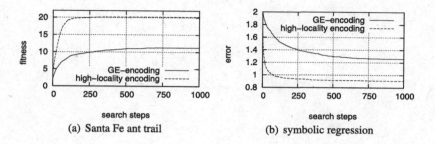

(a) Santa Fe ant trail   **(b) symbolic regression**

Fig. 7.4 Performance of local search using either the GE encoding or a high-locality encoding for the Santa Fe ant trail problem and the symbolic regression problem

The results show that local search using a high-locality representation outperforms local search using the GE representation. Therefore, the low locality of the encoding illustrated in Sect. 7.3 has a negative effect on the performance of modern heuristics. Although the low locality of the GE encoding allows a local search strategy to escape local optima, local search using the GE encoding shows lower performance than a high-locality encoding.

Comparing the results for the two types of encodings reveals that using the GE encoding prolongs search. More search steps are necessary to converge. This increase is expected as for the GE encoding a search step often does not change the corresponding phenotypes. However, the plots show that allowing local search using the GE encoding to run for a higher number of search steps does not increase its performance.

The results show that the GE representation has some problems with locality as neighboring genotypes often do not correspond to neighboring phenotypes. Therefore, a guided search around high-quality solutions can be difficult. However, due to the lower locality of the representation, it is easier to escape from local optima. Comparing local search using either the GE representation or the standard GP encoding (parse trees) with high locality reveals that the low locality of the GE representation reduces the performance of local search.

The results of this experimental case study allow us a better understanding of the functionality of GE and can deliver some explanations for problems of GE that have been observed in the literature. We want to encourage GE researchers to consider locality issues for further developments of the genotype-phenotype mapping. We believe that increasing the locality of the GE representation can also increase the performance and effectiveness of GE.

Chapter 8
Biased Modern Heuristics for the OCST Problem

Biasing modern heuristics is an appropriate possibility in designing problem-specific and high-quality modern heuristics (see Sect. 6.2). If we have knowledge about a problem we can bias the design elements of modern heuristics, namely the representation and search operator, fitness function, the initial solution, or even the search strategy. This chapter presents a case study on how the performance of modern heuristics can be increased by biasing the design elements towards high-quality solutions. Results show that problem-specific and biased modern heuristics outperform standard variants and even for large problem instances high-quality solutions can be found.

The design of optimal communication and transportation networks which satisfy a given set of requirements has been studied extensively in the literature. Many different variants, with or without additional constraints, have been examined, and either exact solutions or heuristics have been given (Kershenbaum, 1993; Chang and Gavish, 1993; Cahn, 1998; Wu and Chao, 2004). An important constrained minimum spanning tree problem is the *optimal communication spanning tree* (OCST) problem which seeks a spanning tree that connects all given nodes and satisfies their communication requirements at a minimum total cost. Like other constraint spanning tree problems, the OCST problem is NP-hard (Garey and Johnson, 1979, ND7). Even medium-sized problem instances are intractable for exact optimization methods and modern heuristics are the methods of choice. The OCST problem is relevant for communication providers, as well as for companies and institutions who want to build up their own communication infrastructure by renting, or buying, communication capacities from communication providers.

Section 8.1 presents and analyzes the OCST problem. We find that optimal solutions are biased towards the minimum spanning tree (MST). Section 8.2 discusses how this result can be used for the design of biased modern heuristics. Consequently, Sect. 8.3 studies the bias of the search operators of the edge-set encoding. The edge-set encoding is a direct representation, whose search operators are biased towards MSTs. In Sect. 8.4, we study the redundant link-biased encoding. We are able to set the encoding-specific parameters such that MST-like solutions are overrepresented. Thus, we can find better solutions in shorter time. Finally, Sect. 8.5 studies how the

F. Rothlauf, *Design of Modern Heuristics*, Natural Computing Series,
DOI 10.1007/978-3-540-72962-4_8, © Springer-Verlag Berlin Heidelberg 2011

performance of modern heuristics depends on the initial solution. Because optimal solutions are similar to MSTs, this intuitive similarity suggests that the performance of modern heuristics can be improved by starting with an MST. Comparing the performance of different types of modern heuristics confirms this conjecture.

8.1 The Optimal Communication Spanning Tree Problem

This section introduces the OCST problem and studies relevant properties which can be used for the design of biased modern heuristics. We start with a problem description. Then, we give a brief overview of optimization methods for OCST problems. As the problem is NP-hard and only constant-factor approximations are possible, no efficient exact approaches are available and modern heuristics are the methods of choice. The section ends with an analysis of the bias of existing OCST test instances. We find that optimal solutions are on average more similar to the minimum spanning tree (MST) than randomly generated solutions.

8.1.1 Problem Description

The *optimal communication spanning tree* (OCST) problem (Hu, 1974), which is also known as the minimum communication spanning tree problem or the simple network design problem (Johnson et al, 1978), seeks a tree that connects all given nodes and satisfies their communication requirements for a minimum total cost. The number and positions of the nodes are given a priori and the cost of a tree is determined by the cost of the edges. An edge's flow is the sum of the communication demands between all pairs of nodes communicating either directly, or indirectly, over the edge. The goal is to find a tree that connects all given nodes and satisfies their communication requirements for a minimum total cost. The cost for each link is not fixed a priori but depends on its length and the flow over this link. Figure 8.1 shows a communication spanning tree on 15 nodes and emphasizes the path connecting nodes 3 and 14.

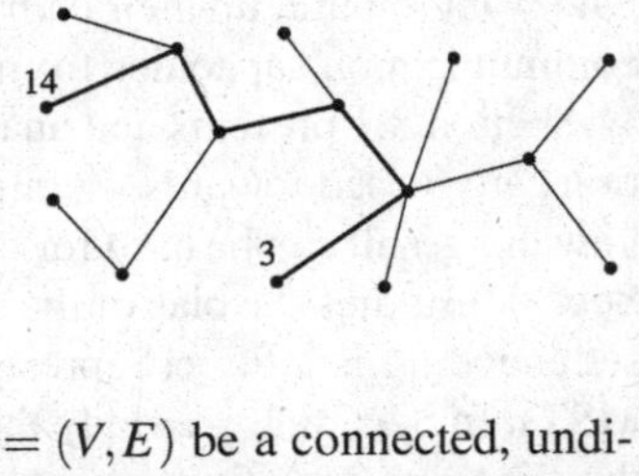

Fig. 8.1 An example for a communication spanning tree on 15 nodes where the path connecting nodes 3 and 14 is emphasized

The OCST problem is defined as follows: Let $G = (V, E)$ be a connected, undirected graph with $n = |V|$ nodes and $m = |E|$ edges. There are *communication or transportation demands* between the n different nodes. An $n \times n$ *demand matrix*

$R = (r_{ij})$ specifies the amount of traffic required between the nodes. An $n \times n$ *distance matrix* $W = (w_{ij})$ specifies the *distance weights* (sometimes also called weights or distances) associated with each pair of nodes. The *weight* $w(T)$ of the spanning tree $T = (V, F)$ $(F \subseteq E)$ is the weighted sum over all pairs of vertices of the cost of the path between all pairs in T. In general,

$$w(T) = \sum_{i,j \in F} f(w_{ij}, tr_{ij}),$$

where the $n \times n$ matrix TR denotes the traffic flowing directly and indirectly over the edge between the nodes i and j. The traffic is calculated according to the demand matrix R and the structure of T. T is a *minimum communication spanning tree* if $w(T) \leq w(T')$ for all other spanning trees T'.

The OCST problem is listed as [ND7] in Garey and Johnson (1979) and Crescenzi and Kann (2003). For the OCST problem as proposed by Hu, the cost f of a link depends linearly on its distance weight w_{ij} and the overall traffic tr_{ij} running over the edge. Therefore, $f = w_{ij} tr_{ij}$ results in the OCST problem

$$w(T) = \sum_{i,j \in F} w_{ij} tr_{ij}, \tag{8.1}$$

which is studied throughout this chapter. In principle, other cost functions f are possible. For example, f can depend non-linearly on its length or traffic, or there are communication lines with only discrete capacities available. The OCST problem becomes the *minimum spanning tree* (MST) problem if $f = w_{ij}$. Then, T is an MST if $w(T) \leq w(T')$ for all other spanning trees T', where $w(T) = \sum_{i,j \in F} w_{ij}$.

Cayley's formula identified the number of spanning trees on n nodes as n^{n-2} (Cayley, 1889). Furthermore, there are n different stars on a tree of n nodes. The dissimilarity between two spanning trees T_i and T_j can be measured using the distance $d_{ij} \in \{0, 1, \ldots, n-2\}$ which is defined as

$$d_{ij} = \frac{1}{2} \sum_{u,v \in V} |e^i_{uv} - e^j_{uv}|. \tag{8.2}$$

e^i_{uv} is 1 if an edge from u to v exists in T_i and 0 if it does not exist in T_i. The number of links that two trees T_i and T_j have in common is then $n - 1 - d_{ij}$. This definition of distance between two trees is based on the Hamming distance (Sect. 2.3.1).

8.1.2 Existing Solution Approaches

Like other constrained spanning tree problems, the OCST problem is NP-hard (Garey and Johnson, 1979, p. 207). Furthermore, Reshef (1999) showed that the problem is APX-hard (Fig. 3.20, p. 90) which means it cannot be solved using a polynomial-time approximation scheme (Sect. 3.4.2), unless P = NP. Therefore, the OCST problem belongs to the class of optimization problems that behave like MAX-SAT (Garey and Johnson, 1979).

Consequently, no efficient algorithmic methods are available. Although some algorithms exist for simplified versions of the OCST problems (*complete unweighted graph problem* (Gomory and Hu, 1961; Hu, 1974) and *uniform demand problem* (Hu, 1974)), there are no efficient methods for standard OCST problems. For detailed information about approximation algorithms for the OCST problem, we refer to the literature (Reshef, 1999; Peleg and Reshef, 1998; Rothlauf, 2009a).

To find high-quality solutions for OCST problems, modern heuristics are the methods of choice (Palmer, 1994; Li, 2001; Rothlauf and Goldberg, 2003; Raidl and Julstrom, 2003; Rothlauf, 2006; Fischer, 2007; Fischer and Merz, 2007). The first modern heuristic was presented by Palmer (1994). He recognized that the design of a proper representation is crucial for the performance of modern heuristics. Palmer compared different representations and developed a new one, the *link and node biased* (LNB) encoding (Sect. 8.4.1). GAs using the LNB encoding showed good results in comparison to a greedy star search heuristic (Palmer, 1994, Chap. 5).

The *characteristic vector* (CV) encoding is a common approach for encoding graphs (Davis et al, 1993; Sinclair, 1995; Berry et al, 1997, 1999) and trees (Berry et al, 1995). It represents a tree or graph as a list of $n(n-1)/2$ binary values. The CV encoding shows good performance when used for trees on a small number of nodes. However, with increasing problem size, the performance of the CV encoding decreases and modern heuristics using this encoding show low performance (compare Rothlauf (2006, Sect. 6.3)).

Weighted encodings, like the LNB encoding (Palmer, 1994), the weighted encoding (Raidl and Julstrom, 2000), the *NetKey* encoding (Rothlauf et al, 2002), or variants of the LNB encoding (Krishnamoorthy and Ernst, 2001) represent a tree using a list of continuous numbers (weights). The weights define an order on the edges, from which the tree is constructed (Sect. 8.4.1). Weighted encodings show good performance when used for tree optimization problems (Rothlauf, 2006). Furthermore, Abuali et al (1995) introduced *determinant factorization*. This representation is based on the in-degree matrix of the original graph, and each factor represents a spanning tree if the determinant corresponding to that factor is equal to one. Tests indicate performance similar to the LNB encoding.

Several modern heuristics using *direct representations* for trees have been presented: a direct representation for trees (Li, 2001), the *edge-set encoding* (Raidl and Julstrom, 2003) (Sect. 8.3.1), and the *NetDir* encoding (Rothlauf, 2006, Chap. 7). The performance results for unbiased direct representations are similar to weighted encodings, however, it is difficult to design search operators in such a way that the search space is uniformly searched (Tzschoppe et al, 2004).

Prüfer numbers were introduced by Prüfer (1918) in a constructive proof of Cayley's theorem (Cayley, 1889), and have subsequently been used to encode spanning trees in modern heuristics (Palmer, 1994; Zhou and Gen, 1997; Gen et al, 1998; Gargano et al, 1998; Gen and Li, 1999; Li et al, 1998; Kim and Gen, 1999; Edelson and Gargano, 2000, 2001). Prüfer numbers belong to the class of *Cayley codes* as they describe a one-to-one mapping between spanning trees on n nodes and strings of $n-2$ node labels. Other Cayley codes have been proposed by Neville (1953) (*Neville II* and *Neville III*), Deo and Micikevicius (2001) (*D-M code*), and Picciotto 1999 (*Blob Code, Happy Code,* and *Dandelion Code*). Caminiti et al (2004) presented a unified approach for Cayley codes which is based on the definition of node pairs, reducing the coding problem to the problem of sorting these pairs into lexicographic order. Due to problems with low locality, Prüfer numbers lead to low performance of modern heuristics (Rothlauf and Goldberg, 1999; Gottlieb et al, 2001; Rothlauf, 2006). Caminiti and Petreschi (2005) showed that the locality of the Blob code is higher than Prüfer numbers, resulting in higher performance of modern heuristics (Julstrom, 2001). Paulden and Smith (2006) extended this work and showed that a single mutation to a Dandelion encoded string leads to at most five edge changes in the corresponding tree, whereas the Prüfer number encoding has no fixed locality bound.

For more details on modern heuristics for OCST problems, we refer to Rothlauf (2006).

8.1.3 Bias of OCST Problems

Test instances for the OCST problem have been proposed by Palmer (1994), Berry et al 1995, and Raidl (2001). Palmer (1994) described OCST problems with six (palmer6), twelve (palmer12), and 24 (palmer24) nodes. The nodes correspond to cities in the US and the distances between the nodes are obtained from a tariff database. The inter-node traffic demands are inversely proportional to the distances between the nodes. Berry et al (1995) presented three OCST test instances, one with six nodes (berry6) and two with 35 nodes (berry35 and berry35u). For berry35u, the distance weights $w_{ij} = 1$. Raidl (2001) proposed various test instances ranging from 10 to 100 nodes. The distance weights and the traffic demands were generated randomly and are uniformly distributed in the interval $[0, 100]$. For all problems the cost of a tree T is $w(T) = \sum_{i,j \in F} w_{ij} tr_{ij}$. Details on the test problems can be found in Rothlauf (2006, Appendix A).

A second set of test instances are random problem instances. The real-valued demands r_{ij} are generated randomly and are usually uniformly distributed either in the interval $[0, 100]$ or $[0, 10]$. The real-valued distance weights w_{ij} are

- generated randomly either in $]0, 100]$ or $[0, 10]$ (random w_{ij}), or
- the Euclidean distances between the nodes placed randomly on a two-dimensional grid of different sizes (e.g. 10×10) (Euclidean w_{ij}).

Again, the cost of a tree T is $w(T) = \sum_{i,j \in F} w_{ij} tr_{ij}$. To measure the performance of modern heuristics, we can use the success probability P_{suc} which is the percentage of runs that find an optimal solution and the number of generations t_{conv} which is the number of generations until a run terminates.

For OCST problems, the cost of a solution depends on the weights w_{ij} of the links used in the tree, since trees that prefer links with low w_{ij} tend to have lower overall costs on average. For Euclidean w_{ij}, usually links near the graph's center of gravity have lower weight than links that are far away from the center of gravity. Therefore, it is useful to run more traffic over the nodes near the center of gravity than over nodes at the edge of the tree (Kershenbaum, 1993). Consequently, nodes can be characterized as either *interior* (some traffic only transits) or *leaf* nodes (all traffic terminates). The more important a link is, and the more transit traffic crosses one of the two nodes it connects, the higher is the degree of the nodes. Nodes near the gravity center tend to have higher degrees than nodes at the edge of tree. Palmer (1994) used this result for the design of the LNB encoding (Sect. 8.4.1). This representation considers the relative importance of nodes in a tree and the more important a node is, the more traffic transits over it (Palmer and Kershenbaum, 1994).

8.1.3.1 Bias of Problem Instances from the Literature

We study whether test instances from the literature have a bias. The goal is to gain knowledge about the problems as well as their optimal solutions. To identify relevant properties of unbiased solutions, we randomly generated 10,000 solutions (trees) for each test problem. Raidl and Julstrom (2003) showed that this is not as simple as it might seem, as techniques based on Prim's or Kruskal's MST algorithms using randomly chosen distance weights do not associate uniform probabilities with spanning trees. An appropriate method for generating unbiased trees is generating random Prüfer numbers (Sect. 8.1.2) and creating the corresponding tree from the Prüfer number (for details on Prüfer numbers, see Rothlauf (2006, Sect. 6.2)).

Table 8.1 Properties of randomly created solutions and optimal solutions for the test instances

problem instance	n	Properties of random solutions				Properties of optimal solutions		
		$d_{mst,rnd}$		$\min(d_{star,rnd})$		$\dfrac{d_{mst,opt}}{n}$	$\dfrac{\min(d_{star,opt})}{n}$	$w(T^{best})$
		$\mu\ (\mu/n)$	σ	$\mu\ (\mu/n)$	σ			
palmer6	6	3.36 (0.56)	0.91	2.04 (0.34)	0.61	1 (0.17)	2 (0.33)	693,180
palmer12	12	9.17 (0.76)	1.17	7.22 (0.60)	0.75	5 (0.42)	7 (0.58)	3,428,509
palmer24	24	21.05 (0.88)	1.30	18.50 (0.77)	0.80	12 (0.50)	17 (0.71)	1,086,656
raidl10	10	7.20 (0.72)	1.10	5.42 (0.54)	0.70	3 (0.30)	4 (0.40)	53,674
raidl20	20	17.07 (0.85)	1.27	14.69 (0.73)	0.77	4 (0.20)	14 (0.70)	157,570
raidl50	50	47.09 (0.94)	1.32	43.88 (0.88)	0.87	13 (0.26)	41 (0.82)	806,864
raidl75	75	72.02 (0.96)	1.36	68.55 (0.91)	0.83	18 (0.24)	68 (0.91)	1,717,491
raidl100	100	97.09 (0.97)	1.36	93.29 (0.93)	0.89	32 (0.32)	90 (0.90)	2,561,543
berry6	6	3.51 (0.59)	0.83	2.03 (0.34)	0.61	0 (0)	2 (0.33)	534
berry35u	35	-	-	29.19 (0.83)	0.83	-	28 (0.80)	16,273
berry35	35	32.05 (0.92)	1.32	29.16 (0.83)	0.83	0 (0)	30 (0.86)	16,915

Table 8.1 lists the properties of randomly created unbiased trees and the properties of the optimal (or best known) solution. It shows the mean μ, the normalized mean μ/n, and the standard deviation σ of the distances $d_{mst,rnd}$ between randomly generated trees and MSTs and of the minimum distances $\min(d_{star,rnd})$ be-

tween randomly created trees and the n different stars. In the instance berry35u, all distance weights are uniform ($w_{ij} = 1$), so all spanning trees are minimal and for each $d_{mst,rnd} = 0$. For the optimal solutions, we calculated their distance to an MST ($d_{mst,opt}$ and $d_{mst,opt}/n$) and their minimum distance to a star ($\min(d_{star,opt})$ and $\min(d_{star,opt})/n$). Furthermore, we list the cost $w(T^{best})$ of optimal solutions. Comparing $d_{mst,opt}$ with $d_{mst,rnd}$ reveals that for all test instances, $d_{mst,opt} < \mu(d_{mst,rnd})$. This means that, on average, the best solution shares more links with an MST than a randomly generated solution. Comparing $\min(d_{star,opt})$ and $\mu(\min(d_{star,rnd}))$ does not reveal any differences. Randomly created solutions have approximately the same expected minimum distance to a star as do the optimal solutions.

8.1.3.2 Bias of Random Problem Instances

For different problem sizes n, we created 100 random problem instances. The demands r_{ij} are generated randomly and are uniformly distributed in $]0, 100]$. For the w_{ij}, there are two possibilities:

- Random: The distance weights w_{ij} are in $]0, 100]$.
- Euclidean: The nodes are randomly placed on a $1,000 \times 1,000$ grid. The w_{ij} are the Euclidean distances between the nodes i and j.

For each of the 100 problem instances, we generated 10,000 random trees.

	$d_{mst,rnd}$		$\min(d_{star,rnd})$	
nodes	μ (μ/n)	σ	μ (μ/n)	σ
8	5.25 (0.66)	1.04	3.74 (0.47)	0.62
12	9.17 (0.77)	1.16	7.31 (0.61)	0.72
16	13.13 (0.82)	1.22	11.00 (0.69)	0.76
20	17.10 (0.86)	1.26	14.78 (0.74)	0.77
24	21.09 (0.88)	1.29	18.60 (0.78)	0.80
28	25.05 (0.895)	1.30	22.70 (0.81)	0.72

Table 8.2 Properties of randomly created solutions for the OCST problem

Table 8.2 presents the mean μ, the normalized mean μ/n, and the standard deviation σ of $d_{mst,rnd}$ and $\min(d_{star,rnd})$. As we get the same results using random or Euclidean distance weights, we neglect the distance weights used. Both the average distance $\mu(d_{mst,rnd})$ between randomly generated solutions and MSTs and the average minimum distance $\mu(\min(d_{star,rnd}))$ between a random solution and a star, increase approximately linearly with n.

For finding optimal or near-optimal solutions for randomly created problem instances, we implement a GA. Although GAs are modern heuristics that cannot guarantee that an optimal solution is found, we choose its design in such a way that an optimal or near-optimal solution is found with high probability. Due to the proposed GA design and due to the NP-hardness of the problem, the effort for finding optimal solutions is high, and optimal solutions can only be determined for small problem instances ($n < 30$). As GA performance increases with population size N, we iteratively apply a GA to the problem and increase N after n_{iter} runs. We start by applying a GA n_{iter} times to an OCST problem using a population size of N_0. T_0^{best} denotes the best solution of cost $w(T_0^{best})$ that is found during the n_{iter} runs. In a next round, we double the population size and again apply the GA n_{iter} times with a population size of $N_1 = 2N_0$. T_1^{best} denotes the best solution with cost $w(T_1^{best})$ that is found in the second round. We continue the iterations and double the population size $N_i = 2N_{i-1}$ until $T_i^{best} = T_{i-1}^{best}$ and $n(T_i^{best})/n_{iter} > 0.5$. This means T_i^{best} is found in more than 50% of the runs in round i. $n(T_i^{best})$ denotes the number of runs that find the best solution T_i^{best} in round i. Although the computational effort of this approach is high and grows exponentially with n, it allows us to find optimal (or near-optimal) solutions for OCST problems with low n.

The GA used is a standard GA (Sect. 5.2) with traditional parameter settings. The problem is encoded using the NetKey encoding (Rothlauf et al, 2002). This representation ensures good GA performance (Rothlauf, 2006) and represents all possible trees uniformly. The standard GA uses one-point crossover and tournament selection without replacement. The size of the tournament is three. The crossover probability is set to $p_{cross} = 0.7$ and the mutation probability (assigning a random value $[0, 1]$ to one allele) is set to $p_{mut} = 1/l$, where $l = n(n-1)/2$.

In our experiments we applied the standard GA for each problem size n to the same 100 problem instances as in Table 8.2. The purpose is to find optimal or near-optimal solutions using a GA and to examine the properties of the best found solutions T_i^{best}. We started with $N_0 = 100$ and set $n_{iter} = 20$. The computational effort for the experiments is high.

Tables 8.3 and 8.4 present the properties of the best solutions T_i^{best} that have been found for the 100 problem instances. We distinguish between Euclidean w_{ij} (Table 8.3) and random w_{ij} (Table 8.4). The tables show for different n the mean μ, the normalized mean μ/n, and the standard deviation σ of $d_{mst,opt}$, $\min(d_{star,opt})$, and $d_{opt,rnd}$, the average population size N_i in the last GA round i, and the mean μ and standard deviation σ of the cost $w(T_i^{best})$ of the best found solution T_i^{best}.

Table 8.3 Properties of optimal solutions for OCST problems with Euclidean distance weights

n	$d_{mst,opt}$ μ (μ/n)	σ	$\min(d_{star,opt})$ μ (μ/n)	σ	$d_{opt,rnd}$ μ (μ/n)	σ	N_i μ	$w(T_i^{best})$ μ	σ
8	1.98 (0.25)	1.18	2.96 (0.37)	0.93	5.25 (0.66)	0.01	389	899,438	177,140
12	4.35 (0.36)	1.47	5.88 (0.49)	1.20	9.17 (0.76)	0.01	1,773	2,204,858	301,507
16	6.74 (0.42)	1.71	9.28 (0.58)	1.24	13.12 (0.82)	0.01	7,048	4,104,579	463,588
20	9.75 (0.49)	1.92	12.68 (0.63)	1.15	17.10 (0.86)	0.01	20,232	6,625,619	649,722
24	11.92 (0.50)	2.14	16.48 (0.69)	1.23	21.08 (0.88)	0.01	40,784	9,320,862	718,694
28	14.60 (0.52)	1.90	19.70 (0.70)	1.42	25.07 (0.89)	0.01	98,467	13,121,110	723,003

The average distance $d_{mst,opt}$ is lower for random w_{ij} than for Euclidean w_{ij}. Comparing the properties of the best found solutions to the properties of randomly created solutions listed in Table 8.2 reveals that $\mu(d_{mst,opt}) < \mu(d_{mst,rnd})$. This means optimal solutions share more links with MSTs than random solutions. Furthermore, the distances $d_{opt,rnd}$ are similar to $d_{mst,rnd}$ and $\min(d_{star,opt})$ is slightly

Table 8.4 Properties of optimal solutions for OCST problems with random distance weights

| | $d_{mst,opt}$ | | $\min(d_{star,opt})$ | | $d_{opt,rnd}$ | | N_i | $w(T_i^{best})$ | |
n	μ (μ/n)	σ	μ (μ/n)	σ	μ (μ/n)	σ	μ	μ	σ
8	0.83 (0.10)	0.88	3.42 (0.43)	0.68	5.25 (0.66)	0.01	234	50,807	18,999
12	1.42 (0.12)	1.04	6.87 (0.57)	0.75	9.16 (0.76)	0.01	478	90,842	28,207
16	2.58 (0.16)	1.36	10.23 (0.64)	0.95	13.12 (0.82)	0.01	2,208	136,275	36,437
20	3.40 (0.17)	1.63	13.94 (0.70)	1.10	17.10 (0.86)	0.01	6,512	183,367	49,179
24	4.08 (0.17)	1.83	17.76 (0.74)	1.04	21.10 (0.88)	0.01	10,432	228,862	58,333
28	5.02 (0.18)	2.07	21.96 (0.78)	0.83	25.06 (0.90)	0.01	19,456	271,897	70,205

(a) distances d over n

(b) normalized distances d/n over n

Fig. 8.2 $\mu(d_{mst,rnd})$ and $\mu(d_{mst,opt})$ over n for random and Euclidean distance weights. The error bars indicate the standard deviation

lower than $\min(d_{star,rnd})$, especially for Euclidean w_{ij}. However, this effect is weak and can be neglected in comparison to the low distance $d_{mst,opt}$.

Figures 8.2(a) and 8.2(b) summarize the results from Tables 8.2, 8.3, and 8.4. They plot $d_{mst,rnd}$ ($d_{mst,rnd}/n$) and $d_{mst,opt}$ ($d_{mst,opt}/n$) using either random w_{ij} or Euclidean w_{ij} over n. Figure 8.2(a) indicates that all distances to the MST increase approximately linearly with n and that optimal solutions share many edges with MSTs. Finally, Fig. 8.3 shows the distribution of $d_{mst,opt}$ for Euclidean and random w_{ij} for the 100 randomly generated instances of 20-node OCST problems as well as the distribution of $d_{mst,rnd}$ for 100 randomly generated trees. We see that optimal solutions share many edges with MSTs.

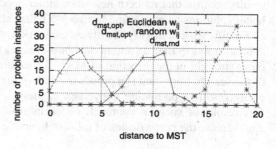

Fig. 8.3 Distribution of distance $d_{mst,opt}$ between MST and optimal solutions (Euclidean and random w_{ij}) and distance $d_{mst,rnd}$ between MST and random solution. Results are for OCST problems with $n = 20$

8.2 Biasing Modern Heuristics for OCST Problems

The similarity between optimal solutions and MSTs can be used for biasing modern heuristics. Relevant design elements are the problem representation and search operators, the initialization method, and the fitness function. Consequently, four promising approaches on how to consider the similarity between optimal solutions and MSTs can be identified:

- Design search operators that favor trees similar to an MST.
- Design a redundant problem representation such that solutions similar to an MST are overrepresented.
- Start with solutions that are similar to an MST.
- Assign a higher fitness value to solutions similar to an MST.

The following paragraphs discuss these four possibilities. We can bias search operators to favor MST-like trees. If this principle is applied to OCST problems, search operators will prefer edges of low weight. This concept was implicitly used by Raidl and Julstrom (2003) who proposed the edge-set representation. The search operators of edge-sets can be combined with additional heuristics (Sect. 6.2.1) to prefer low-weight edges. As a result, the search operators favor MST-like solutions which can result in higher performance in comparison to unbiased search operators (Raidl and Julstrom, 2003). In Sect. 8.3, we study how the performance of modern heuristics using edge-sets depends on the properties of optimal solutions.

For redundant representations, the number of genotypes exceeds the number of phenotypes. Redundant representations can increase the performance of modern heuristics if their bias appropriately exploits an existing bias of optimal solutions (Sect. 6.2.2). Therefore, biasing a redundant representation towards MSTs can increase the performance of modern heuristics for OCST problems. This principle is, for example, used in the link-biased encoding (Palmer, 1994). In Sect. 8.4, we study this redundant representation and examine how it influences the performance of modern heuristics.

The performance of modern heuristics depends on the starting point of the search. Therefore, as the average distance between optimal solutions and MSTs is low, it is a reasonable approach to start the search with solutions that are an MST or similar to it. We make use of this observation in Sect. 8.5 and examine how the performance of different modern heuristics depends on the starting solution. The experimental results confirm that modern heuristics either need fewer fitness evaluations or find better solutions when starting from a MST than starting from a random tree.

Finally, we can modify the fitness evaluation of trees such that solutions that are similar to an MST get an additional bonus. Such a bias could push modern heuristics more strongly in the direction of MST-like solutions and increase the performance of the search. Important is a proper strength of the bias since a too strong bias could easily result in premature convergence. Steitz and Rothlauf (2010) presented the first results of such an approach which indicate that biasing the fitness function can increase the performance of modern heuristics.

When designing problem-specific modern heuristics, we must have in mind that favoring MST-like solutions is only reasonable for problems where optimal solutions are similar to MSTs. Section 8.1.3 presents experimental evidence that OCST problems have this specific property. However, if there is no such problem-specific knowledge (e.g. all distance weights $w_{ij} = 1$ and all trees are MSTs), it is not reasonable to use the proposed techniques. Therefore, methods that can solve OCST problems efficiently are not necessarily appropriate for other tree problems where optimal solutions are not MST-like. Using OCST-specific optimization methods for tree problems with unknown properties can result in low performance.

8.3 Search Operator

When using modern heuristics for tree problems, it is necessary to encode a solution (tree) such that search operators can be applied. There are two different possibilities for doing this: indirect representations usually encode a tree (phenotype) as a list of strings (genotypes) and apply standard search operators to the genotypes. The phenotype is constructed by an appropriate genotype-phenotype mapping (representation). In contrast, direct representations encode a tree as a set of edges and apply search operators directly to this set. Therefore, no representation is necessary. Instead, tree-specific search operators must be developed as standard search operators can no longer be used (Sect. 4.3.4).

The *edge-set encoding* (Raidl and Julstrom, 2003) is a representative example of a direct representation. Raidl and Julstrom (2003) proposed two different variants: heuristic variants where the search operators consider the weights of the edges, and non-heuristic ones. Results from applying the edge-set encoding to two sets of degree-constrained MST problem instances have indicated the superiority of edge-sets in comparison to several other codings of spanning trees (i.e. the Blob Code, network random keys, and strings of weights) particularly when the operators implement edge-cost-based heuristics (Raidl and Julstrom, 2003, p. 238).

In this section, we investigate the bias of the search operators of edge-sets, and study how the performance of modern heuristics depends on the properties of optimal solutions. As the heuristic variants of edge-sets prefer edges with a low cost, these variants are expected to show a bias towards MSTs (Rothlauf, 2009b).

In Sect. 8.3.1, we introduce the functionality of the edge-set encoding with and without heuristics. Section 8.3.2 studies the bias of the search operators. Finally, Sect. 8.3.3 examines the performance of modern heuristics using edge-sets for OCST problems. It presents results for known test instances from the literature as well as randomly generated test instances.

8.3.1 Search Operators of the Edge-Set Encoding

The edge-set encoding directly represents trees as sets of edges. Therefore, encoding-specific initialization, recombination, and local search operators are necessary. The following paragraph summarizes the functionality of the different variants with and without heuristics (Raidl and Julstrom, 2003).

8.3.1.1 Edge-Set Encoding without Heuristics

Raidl and Julstrom (2003) proposed and investigated three different initialization strategies: PrimRST, RandWalkRST, and KruskalRST. PrimRST, which is based on Prim's algorithm (Prim, 1957), slightly overrepresents stars and underrepresents lists. RandWalkRST has an average running time of $O(n \log n)$, however, the worst-case running time is unbounded. Therefore, Raidl and Julstrom (2003) recommended the use of the KruskalRST which is based on the algorithm of Kruskal (Kruskal, 1956). In contrast to Kruskal's algorithm, KruskalRST chooses edges (i, j) not according to their corresponding weights w_{ij} but randomly. KruskalRST has a small bias towards star-like trees (which is lower than the bias of PrimRST). The bias is a result of the sequential insertion of edges during the construction of a tree using Kruskal's algorithm. Algorithm 9 outlines the functionality of KruskalRST.

Algorithm 9 KruskalRST(V, E)

$T \leftarrow \emptyset, A \leftarrow E;$ //E is the set of available edges
while $|T| < |V| - 1$ **do**
   choose an edge $\{(u, v)\} \in A$ at random;
   $A \leftarrow A - \{(u, v)\};$
   if u and v are not yet connected in T **then**
      $T \leftarrow T \cup \{(u, v)\};$
   end if
end while
return T

To perform recombination and obtain an offspring T_{off} from two parental trees T_1 and T_2 with edge sets E_1 and E_2, KruskalRST is applied to the graph $G_{cr} = (V, E_1 \cup E_2)$. Instead of KruskalRST, in principle PrimRST and RandWalkRST can also be used. The recombination operator has high *heritability* (Caminiti and Petreschi, 2005; Raidl and Gottlieb, 2005) as in the absence of constraints only parental edges are used to create the offspring. Raidl and Julstrom (2003) distinguished two different recombination operators: the variant previously described is denoted KruskalRST crossover. The second variant is denoted KruskalRST* crossover. When using this variant, in a first step all shared dges $(E_1 \cap E_2)$ are included in the offspring T_{off}. Then T_{off} is completed by applying KruskalRST to the remaining edges $(E_1 \cup E_2) \setminus (E_1 \cap E_2)$. Results from Raidl and Julstrom (2003) indicate a better performance of KruskalRST* for the degree-constraint MST problem.

The local search operator (also called mutation) randomly replaces one edge in the spanning tree. This replacement can be realized in two different ways. The first variant randomly chooses one edge that is not present in T and includes it in T. Then, a randomly chosen edge of the cycle is removed ("insertion before deletion"). The second variant first randomly deletes one edge from T and then connects the two disjoint connected components using a random edge not present in T ("deletion before insertion"). The running time is $O(n)$.

8.3.1.2 Edge-Set Encoding with Heuristics

We describe how heuristics that rely on the weights w_{ij} can be included in the edge-set encoding. Raidl and Julstrom (2003) introduced these variants of edge-sets due to the assumption that in weighted tree optimization problems optimal solutions often prefer edges with low weights w_{ij}.

To favor low-weighted edges when generating an initial population, the algorithm KruskalRST does not choose an edge at random but sorts all edges in the underlying graph according to their weights w_{ij} in ascending order. The first spanning tree in a population is created by choosing the first edges in the ordered list. As these are the edges with lowest weights, the first generated spanning tree is an MST. Then, the k edges with lowest weights are permuted randomly and another tree is created using the first edges in the list. This heuristic initialization results in a strong bias towards MSTs. With increasing k, the bias of randomly created trees towards MSTs is reduced. The number of edges, which are permuted increases according to

$$k = \alpha(i-1)n/N$$

where N denotes the population size, i is the number of the tree that is actually generated ($i = 1, \dots, N$) and α, with $0 \le \alpha \le (n-1)/2$, is a parameter that controls the strength of the heuristic bias.

The heuristic recombination operator is a modified version of KruskalRST* crossover. Firstly, the operator transfers all edges $(E_1 \cap E_2)$ that exist in both parents T_1 and T_2 to the offspring. Then, the remaining edges are chosen randomly from $E' = (E_1 \cup E_2) \setminus (E_1 \cap E_2)$ using a tournament with replacement of size two. This means the weights w_{ij} of two randomly chosen edges are compared and the edge with the lower weight is inserted into the offspring (if no cycle is created).

The heuristic mutation operator is based on mutation by "insertion before deletion". In a pre-processing step, all edges in the underlying graph are sorted according to their weights in ascending order. Doing this, a rank is assigned to every edge. Rank one is assigned to the edge with lowest weight. To favor low-weighted edges, the edge that is inserted by the heuristic mutation operator is not chosen randomly but according to its rank

$$R = \lfloor |\mathcal{N}(0, \beta n)| \rfloor \bmod m + 1,$$

where $\mathcal{N}(0, \beta n)$ is the normal distribution with mean 0 and standard deviation βn and $m = n(n-1)/2$. β is a parameter that controls the bias towards low-weighted edges. If a chosen edge already exists in T, the edge is discarded and the selection is repeated.

8.3.2 Bias

We investigate the bias of the initialization, mutation, and recombination operators. We present results for random and Euclidean test problems (Sect. 8.1.3).

8.3.2.1 Initialization

Raidl and Julstrom (2003) examined the bias of different initialization methods and found KruskalRST to be slightly biased towards stars. As the bias is sufficiently small and due to its lower running time it is preferred in comparison to RandWalkRST and PrimRST, which shows a stronger bias towards stars.

Table 8.5 shows the average distances $d_{mst,rnd}$ between MSTs and randomly generated trees T_{rand} (the standard deviations are shown in brackets). For each problem instance (250 of each type) we generated 1,000 random solutions T_{rand} using either an unbiased encoding (Prüfer numbers), non-heuristic KruskalRST (Sect. 8.3.1.1), or the heuristic initialization (Sect. 8.3.1.2). For the heuristic initialization α was set either to $\alpha = 1.5$ as recommended in Raidl and Julstrom (2003) or to the maximum value $\alpha = (n-1)/2$, which results in the lowest possible bias. The results confirm that KruskalRST is not biased towards MSTs (Raidl and Julstrom, 2003). Furthermore, the heuristic versions show a bias towards MSTs even when using a large value of α.

Table 8.5 Mean and standard deviation of distances $d_{mst,rnd}$ between random trees and MST

n	weights	unbiased	KruskalRST non-heuristic	heuristic initialization	
				$\alpha = 1.5$	$\alpha = (n-1)/2$
10	Euclidean	7.20 (0.06)	7.20 (0.04)	0.44 (0.19)	3.80 (0.08)
	random		7.20 (0.07)	0.20 (0.13)	3.74 (0.08)
20	Euclidean	17.10 (0.04)	17.10 (0.04)	1.06 (0.28)	12.02 (0.14)
	random		17.10 (0.04)	0.42 (0.22)	12.09 (0.08)
100	Euclidean	97.02 (0.04)	97.02 (0.04)	5.98 (0.67)	87.89 (0.18)
	random		97.02 (0.05)	2.12 (0.60)	88.22 (0.87)
200	Euclidean	197.03 (0.04)	197.02 (0.04)	11.99 (0.92)	176.45 (0.29)
	random		197.00 (0.04)	3.93 (0.69)	177.25 (0.12)

8.3.2.2 Recombination

To investigate whether the crossover operator of the edge-set encoding leads to an over-representation of MST-like individuals, we randomly generate an initial population of $N = 50$ individuals. Then, in each search step, one random offspring is generated which replaces one randomly chosen individual of the population. The offspring is created either by recombination alone (with probability $p_c = 1$) or by recombination ($p_c = 1$) and mutation (with mutation probability $p_m = 1/n$). We present results for the following combinations of search operators:

- **nohxover only**: non-heuristic KruskalRST* crossover with $p_c = 1$, no mutation ($p_m = 0$), and non-heuristic initialization with KruskalRST.
- **hxover only**: heuristic crossover with $p_c = 1$, no mutation ($p_m = 0$), and non-heuristic initialization (KruskalRST).
- **hxover, nohmut**: heuristic crossover ($p_c = 1$), non-heuristic mutation with $p_m = 1/n$ and "insertion before deletion", and non-heuristic initialization (KruskalRST).
- **hxover, hmut** ($\beta = 1$): heuristic crossover ($p_c = 1$), heuristic mutation ($p_m = 1/n$) with $\beta = 1$, and non-heuristic initialization (KruskalRST).
- **hxover, hmut** ($\beta = 5$): heuristic crossover ($p_c = 1$), heuristic mutation ($p = 1/n$) with $\beta = 5$, and non-heuristic initialization (KruskalRST).
- **hxover, hmut** ($\beta = 1$), **hini**: heuristic crossover ($p_c = 1$), heuristic mutation ($p = 1/n$) with $\beta = 1$, and heuristic initialization with $\alpha = 1.5$.
- **hxover, hmut** ($\beta = 5$), **hini**: heuristic crossover ($p_c = 1$), heuristic mutation ($p = 1/n$) with $\beta = 5$, and heuristic initialization with $\alpha = 1.5$.

We perform experiments for different problem sizes n and perform $iter = 5,000$ ($n = 25$) or $iter = 10,000$ ($n = 100$) search steps. As no selection operator (Sect. 5.2.1) is used, no selection pressure pushes the population towards high-quality solutions. Search operators are unbiased if the statistical properties of the population do not change by applying variation operators alone. In our experiments we measure in each generation the average distance $d_{mst-pop} = (1/N) \sum_{i=1}^{N} d_{i,mst}$ between all individuals T_i ($i \in \{0, \ldots, N-1\}$) in the population and an MST. If $d_{mst-pop}$ decreases, the variation operators are biased towards MSTs. If $d_{mst-pop}$ remains constant, the variation operators are unbiased with respect to MSTs.

As for initialization, we perform this experiment on 250 randomly generated tree instances of different sizes with random distance weights w_{ij}. Results for Euclidean distance weights are analogous. For every tree instance, we perform 10 runs with different randomly chosen initial populations.

Figure 8.4 shows the mean of $d_{mst-pop}$ over the number of search steps. For clarity, standard deviations are generally omitted; exemplarily we plot the standard deviations for "nohxover only". The results confirm previous findings (Tzschoppe et al, 2004) and show that the crossover operator without heuristics is not biased towards MSTs and does not modify the statistical properties of the population ($d_{mst-pop}$ remains constant over the number of search steps). Furthermore, heuristic crossover operators show a strong bias towards MSTs. Applying heuristic crossover operators alone (hxover only) pushes the population towards MSTs. After some search steps

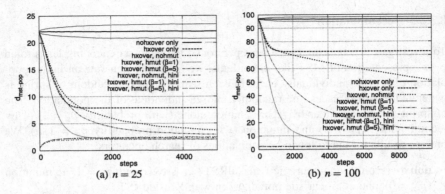

Fig. 8.4 Mean of the distance $d_{mst-pop}$ between a population of $N = 50$ individuals and an MST over the number of search steps

(for example, after $1,000$ search steps for $n = 25$ and $N = 50$), the population has fully converged and crossover-based search gets stuck.

The problem of premature convergence can be amended when combining heuristic crossover with heuristic or non-heuristic mutation. Then, new edges are continually introduced into the population. Consequently, with an increasing number of search steps, the population keeps moving towards an MST. However, in contrast to when using no mutation, the population does not reach an MST as the mutation operator continuously inserts new edges (hxover, nohmut). The situation is similar when replacing the non-heuristic mutation operator by heuristic variants with $\beta = 1$ or $\beta = 5$ (hxover, hmut ($\beta = 1$) or hxover, hmut ($\beta = 5$)). With lower β, the bias towards low-weighted edges increases and the population converges faster towards MSTs. Similarly to when using non-heuristic mutation, the population does not fully converge to MSTs as the mutation operator (heuristic as well as non-heuristic) continually inserts new edges into the population.

Table 8.5 indicates that heuristic initialization is biased towards MSTs. When using heuristic initialization with $\alpha = 1.5$, the plots ("hxover, nohmut, hini", "hxover, hmut ($\beta = 1$), hini", and "hxover, hmut ($\beta = 5$), hini") indicate that the variation operators allow the population to recover from the strong initial bias. With an increasing number of search steps, $d_{mst-pop}$ converges to the same values as when using non-heuristic initialization.

Summarizing the results, non-heuristic crossover results in an unbiased population. When using the heuristic crossover operator, the population converges to solutions only slightly different from an MST.

8.3.2.3 Mutation

Finally, we investigate the bias of the mutation (local search) operator for 250 randomly generated tree instances with random distance weights w_{ij}. As for the crossover operator, we create a random population of $N = 50$ individuals. Then, in every search step, a randomly chosen individual is mutated exactly once using either the non-heuristic "insertion-before-deletion" mutation or the heuristic version. The mutated offspring replaces a randomly selected individual in the population. Therefore, no intensification mechanisms focus the search. For the heuristic mutation operator the parameter β is set to 1, 5, or 10. With lower β, edges with lower weights are preferred. The initial population is generated randomly using either the non-heuristic initialization or the heuristic variant with $\alpha = 1.5$ (Raidl and Julstrom, 2003). We perform experiments for $n = 25$ ($iter = 5,000$) and $n = 100$ ($iter = 10,000$).

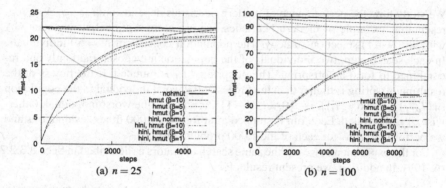

(a) $n = 25$ (b) $n = 100$

Fig. 8.5 Mean of the distance $d_{mst-pop}$ between a population and MSTs over the number of search steps

Figure 8.5 shows the mean of $d_{mst-pop}$ over the number of search steps. The results show that the non-heuristic mutation operator is approximately unbiased, whereas the heuristic mutation is biased towards MSTs. As expected, the bias increases with lower β. Furthermore, the population does not converge completely towards MSTs but the average distance $d_{mst-pop}$ remains stable after a number of search steps. For heuristic initialization with $\alpha = 1.5$, the initial population shows a strong bias towards MSTs (see Table 8.5, p. 198) but the population can recover due to the mutation operator. With increasing number of search steps, the population converges to the same $d_{mst-pop}$ as when using the non-heuristic initialization.

8.3.3 Performance Results

To be able to study how the performance of modern heuristics using edge-sets depends on the structure of the optimal solution T_{opt}, an optimal or near-optimal solution must be determined. However, due to the NP-hardness of the OCST problem, guaranteed optimal solutions can only be determined with reasonable computational effort for small problem instances. Therefore, we split our study into three parts: the first part is concerned with small 20 node problem instances where we are able to determine optimal (or at least near-optimal) solutions using the GA framework described in Sect. 8.1.3.2. The second part deals with known test problems from the literature, and the third part studies larger problems with unknown T_{opt}.

8.3.3.1 Influence of $d_{mst,opt}$

We study how the performance of modern heuristics depends on $d_{mst,opt}$. As a representative example of modern heuristics, we use a conventional steady-state GA with an $(N + 1)$-selection strategy. The population consists of $N = 50$ individuals. In each search step, two individuals of the population are chosen randomly and recombined to form an offspring. The offspring T_{off} is mutated. If the fitness of the resulting offspring is higher than the fitness of the worst individual in the population $(w(T_{off}) \leq \max(w(T_i)), i \in \{0, \ldots, N-1\})$, it replaces the worst individual. Otherwise, it is discarded. Each run is stopped after $iter = 10,000$ fitness evaluations and we perform 10 runs for each of the 1,000 problem instances.

For the experiments, we use the same search operators as described in Sect. 8.3.2.2 (p. 199). In addition, we present results for

- **nohxover, nohmut**: non-heuristic KruskalRST* crossover $(p_c = 1)$, non-heuristic mutation $(p_m = 1/n)$, and non-heuristic initialization (KruskalRST).

Results for $n = 20$ are presented in Fig. 8.6. We plot the percentage of runs that find the optimal solution T_{opt} over $d_{mst,opt}$. Results are averaged over all 1,000 randomly generated problem instances. The results reveal that GAs using heuristic crossover always find the optimal solution T_{opt} if it is similar to an MST ($d_{mst,opt}$ is low, Fig. 8.6(a)). For larger $d_{mst,opt}$, GA performance drops sharply and the percentage of runs that find T_{opt} becomes low. Using heuristic initialization results in lower GA performance for larger $d_{mst,opt}$ in comparison to using non-heuristic initialization. This effect is due to the strong bias of the heuristic initialization towards MSTs.

In contrast to the heuristic crossover operator, the performance of GAs using the non-heuristic KruskalRST* operator ("nohxover, nohmut") decreases only slightly with larger $d_{mst,opt}$ and allows GAs to find optimal solutions even for larger $d_{mst,opt}$. Combining the non-heuristic crossover with heuristic mutation (nohxover, hmut ($\beta = 1$)) results in higher performance than when using non-heuristic mutation (nohxover, nohmut) if $d_{mst,opt}$ is low (Fig. 8.6(a)). This is a result of the bias of the heuristic mutation operator towards MSTs. Analogously, if $d_{mst,opt}$ is large (Fig. 8.6(b)), the use of non-heuristic mutation (nohxover, nohmut) results in higher

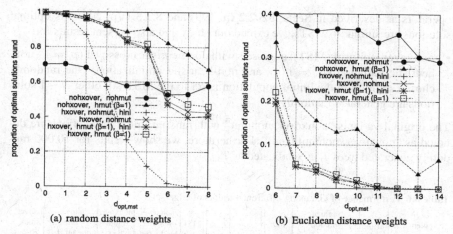

Fig. 8.6 Performance of a steady-state $(50+1)$-GA using different variants of the edge-set encoding for randomly generated $n = 20$ node OCST problems

performance than the use of biased heuristic mutation (nohxover, hmut ($\beta = 1$)). Furthermore, OCST problems with random w_{ij} are easier to solve for GAs using heuristic variants of edge-sets since optimal solutions are more similar to MSTs.

In summary, the heuristic variants of the edge-set encoding only perform well for OCST problems where T_{opt} is only slightly different from an MST. Otherwise, EAs using the heuristic crossover operator fail due to their strong bias towards MSTs. Using heuristic mutation has a similar, but weaker effect on GA performance than heuristic crossover. If $d_{mst,opt}$ is low, its bias towards MSTs helps GAs to find optimal solutions; in contrast, if $d_{mst,opt}$ is large, GA performance is reduced. The performance of GAs using non-heuristic crossover and mutation operators is nearly independent of $d_{mst,opt}$. Therefore, also problems with large $d_{mst,opt}$ can be solved.

8.3.3.2 Test Instances from the Literature

We examine GA performance for the OCST test instances from Sect. 8.1.3. Since for all test instances optimal or near-optimal solutions are known, we can also study how GA performance depends on $d_{mst,opt}$. Table 8.6 lists the properties of optimal or best known solutions T_{opt} for the test instances. It shows the number of nodes n, the average distance $d_{mst,rnd}$ of 10,000 randomly generated unbiased trees T_{rand}, the distance $d_{mst,opt}$, and the cost $w(T_{opt})$. For all test instances, $d_{mst,opt}$ is always smaller than $d_{mst,rnd}$.

For our experiments we use the same steady-state GA as described in the previous paragraphs. Each run is stopped after $iter$ fitness evaluations and we perform 25 runs for each test instance. We compare the performance of GAs that use the same search operators as described in Sects. 8.3.2.2 (p. 199) and 8.3.3.1 (p. 202). As optimal solutions are similar to MSTs, we extend our study and also present results for:

- **nohxover, nohmut (MST)**: edge-sets with non-heuristic crossover ($p_c = 1$), non-heuristic mutation ($p_m = 1/n$), and non-heuristic initialization. One randomly chosen solution in the initial population is set to the MST (all other $N - 1$ solutions are generated randomly).

The original variants "nohxover, nohmut" with random initial populations are denoted as "nohxover, nohmut (rnd)". Furthermore, we show results for 10,000 randomly generated trees T_{rand} (indicated as T_{rand}).

Table 8.6 Performance of GA using different variants of edge-sets

problem instance	n	$d_{mst,rnd}$	optimal solution $d_{mst,opt}$	optimal solution $w(T_{opt})$	iter		T_{rand}	MST	nohxover nohmut MST	nohxover nohmut rnd	nohxover hmut $\beta=1$	hxover, hini nohmut	hxover, hini $\beta=1$	hxover hmut, $\beta=1$
palmer6	6	3.36	1	693,180	300	P_{suc}	-	0	0.64	0.64	0.82	0.98	1	1
						$d_{bf,opt}$	-	1	0.52	0.54	0.26	0.04	0	0
						gap (%)	149	2.39	0.52	0.68	0.30	0.04	0	0
palmer12	12	9.17	5	3,428,509	5,000	P_{suc}	-	0	0.2	0.26	0.24	0	0.04	0.12
						$d_{bf,opt}$	-	5	2.72	2.62	2.06	3.08	1.56	1.46
						gap (%)	244	13.07	1.07	1.02	0.61	1.74	0.47	0.40
palmer24	24	21.05	12	1,086,656	20,000	P_{suc}	-	0	1	1	0	0.88	0	0
						$d_{bf,opt}$	-	12	0	0	3.2	0.12	6.48	4.84
						gap (%)	852	80.35	0	0	5.93	0.01	29.68	22.35
raidl10	10	7.20	3	53,674	1,000	P_{suc}	-	0	0.58	0.48	1	0	0.5	0.74
						$d_{bf,opt}$	-	3	0.44	0.64	0	1.6	0.5	0.26
						gap (%)	512	8.72	0.94	1.76	0	3.25	0.89	0.46
raidl20	20	17.07	2	157,570	5,000	P_{suc}	-	0	0.44	0	0.74	0	1	0.96
						$d_{bf,opt}$	-	2	0.76	6.8	1.12	1.18	0	0.24
						gap (%)	1,145	5.22	1.10	33.13	0.86	0.59	0	0.18
raidl50	50	47.09	13	806,864	40,000	P_{suc}	-	0	0.067	0	0.467	0	0	0
						$d_{bf,opt}$	-	13	4.27	12.53	1.27	8.73	5.53	4.13
						gap (%)	2,493	13.07	0.80	7.98	0.15	4.10	1.59	1.10
raidl75	75	72.02	18	1,717,491	40,000	P_{suc}	-	0	0	0	0.067	0	0	0
						$d_{bf,opt}$	-	18	12.4	46.33	12.27	14.8	2.87	3.87
						gap (%)	3,414	39.95	4.64	57.80	3.00	7.61	0.26	0.66
raidl100	100	97.09	32	2,561,543	80,000	P_{suc}	-	0	0	0	0.267	0	0	0
						$d_{bf,opt}$	-	32	20	53.53	7.87	22.47	2.2	4.67
						gap (%)	4,534	40.82	5.80	37.01	1.20	7.32	0.53	0.58
berry6	6	3.51	0	534	300	P_{suc}	-	1	1	0.88	1	1	1	1
						$d_{bf,opt}$	-	0	0	0.12	0	0	0	0
						gap (%)	140	0	0	0.57	0	0	0	0
berry35u	35	-	-	16,273	40,000	P_{suc}	-	0	-	0.52	0.02	0.02	0	0.02
						$d_{bf,opt}$	-	-	6.64	6.1	12.44	6.6	13.82	11.86
						gap (%)	308	37.22	2.19	1.54	8.31	2.13	10.16	6.92
berry35	35	32.05	0	16,915	20,000	P_{suc}	-	1	1	0.92	1	1	1	1
						$d_{bf,opt}$	-	0	1	0.46	0.64	1	1	0.96
						gap (%)	2,143	0	0	0.23	0	0	0	0

Table 8.6 lists the percentage of runs P_{suc} that find T_{opt}, the average distance $d_{bf,opt}$ between the best solution T_{bf} that was found after *iter* fitness evaluations and T_{opt}, and the average gap $\frac{w(T_{bf})-w(T_{opt})}{w(T_{opt})}$ (in percent) between $w(T_{bf})$ and $w(T_{opt})$.

For the 10,000 randomly generated solutions T_{rand}, we present the average gap $\frac{w(T_{rand})-w(T_{opt})}{w(T_{opt})}$ (in percent).

We see that the gap between $w(T_{mst})$ and $w(T_{opt})$ is much lower than the gap between $w(T_{rand})$ and $w(T_{opt})$. Therefore, an MST is always already a high-quality solution for all test problems. For berry6 and berry35, where an MST is optimal, GAs using heuristic crossover (hxover) always and easily find T_{opt}. For test instances with a large $d_{mst,opt}$ (e.g. the large Raidl test instances), GAs using heuristic crossover have problems finding T_{opt}. Although, the gap between $w(T_{bf})$ and $w(T_{opt})$ is low and similar to GAs using non-heuristic crossover and heuristic mutation (nohxover, hmut ($\beta = 1$)), the strong bias of the heuristic crossover operator does not allow the GA to reach T_{opt}.

High GA performance is obtained (except for palmer24) when combining non-heuristic crossover with heuristic mutation (nohxover, hmut ($\beta = 1$)). Combining heuristic mutation with the unbiased non-heuristic crossover operator seems to be a good compromise and also allows high GA performance for problems where $d_{mst,opt}$ is large.

The results confirm the findings from the previous paragraphs. The non-heuristic KruskalRST* crossover shows good performance also for large $d_{mst,opt}$. The heuristic crossover operator only finds optimal solutions if they are similar to an MST. However, the gap between $w(T_{bf})$ and $w(T_{opt})$ is low as good solutions are often MST-like.

8.3.3.3 Random Test Instances

Finally, we study the GA performance for randomly generated, large OCST problem instances. In contrast to the previous paragraphs, we do not know T_{opt} and GA performance is determined only by the fitness $w(T_{bf})$. We must bear in mind that optimal solutions are similar to MSTs, and in comparison to randomly chosen trees, an MST is already a high-quality solution.

As before, we use a steady-state GA with $N = 50$ which is stopped after *iter* fitness evaluations (Table 8.7). We present results for randomly generated OCST problems of different sizes ($n = 10$, $n = 25$, $n = 50$, $n = 100$, $n = 150$, $n = 200$) with either random or Euclidean distance weights w_{ij}. For each type of problem, we create either 100 ($n < 100$), 25 ($100 \leq n \leq 150$), or 10 ($n = 200$) random problems. For each random problem, 10 GA runs are performed. We compare GA performance using the search operators described in Sects. 8.3.2.2 (p. 199) and 8.3.3.2 (p. 204).

Table 8.7 Number of fitness evaluations *iter* for large OCST problems

n	10	25	50	100	150	200
iter	1,000	5,000	20,000	80,000	160,000	320,000

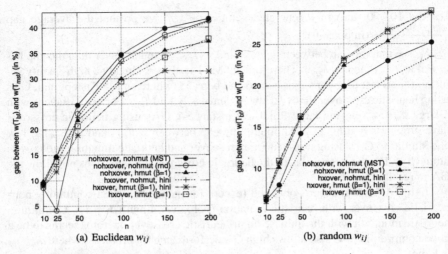

(a) Euclidean w_{ij} (b) random w_{ij}

Fig. 8.7 GA performance for large problem instances

Figure 8.7 presents the average gap $\frac{w(T_{mst})-w(T_{bf})}{w(T_{mst})}$ (in percent). The higher the gap, the better the solutions found. We compare the performance between different search operators based on the fitness gap between T_{bf} and MST since an MST is already a high-quality solution and since the design of the search operators used is such that MSTs (or slight variants of it) are created with a high probability in either the initial population (heuristic initialization), or during the GA run (heuristic mutation or crossover). For random distance weights, we do not present results for "nohxover, nohmut (rnd)" as GAs using this representation were unable to find solutions that have a similar or lower cost than the MST.

The results show differences between Euclidean and random problem instances. For Euclidean w_{ij}, best solutions are found when using the non-heuristic variant of edge-sets starting with an MST ("nohxover, nohmut (MST)"). In contrast, solution quality is low for heuristic initialization, crossover, and mutation (hxover, hmut ($\beta = 1$), hini). The situation is reversed for random w_{ij}. GA performance is high for heuristic initialization, crossover, and mutation; in contrast, GA performance is lower when using the non-heuristic variants. This is because random w_{ij} result in a stronger bias of optimal solutions towards the MST (Fig. 8.3). Therefore, encodings with a strong bias towards MSTs result in higher GA performance if optimal solutions are similar to MSTs, but lower GA performance otherwise. Consequently, encodings like "hxover, hmut ($\beta = 1$), hini" with a high bias perform better for random w_{ij} (optimal solutions are more similar to MSTs) than for Euclidean w_{ij}.

The results confirm previous findings. Heuristic edge-sets result in high GA performance if optimal solutions are similar to MSTs (random weights). In contrast, for larger $d_{mst,opt}$ (Euclidean weights) non-heuristic variants show better performance as the bias of the heuristic variants hinders finding optimal solutions.

8.4 Representation

This section examines the *link-biased* (LB) encoding, which is a (indirect) representation for trees. The LB encoding is a redundant representation, which can encode trees similar to an MST with higher probability than random trees. In Sect. 8.1.3, we found that optimal solutions of OCST problems are MST-like. The LB encoding exploits this problem-specific knowledge. Experimental results show high performance for different types of modern heuristics. This section illustrates how biasing a redundant representation towards high-quality solutions can increase the performance of modern heuristics (Rothlauf, 2009c).

Section 8.4.1 describes the functionality of the LB encoding. Section 8.4.2 presents experimental results on a proper setting of the encoding-specific parameter P_1 and the performance of the LB encoding for various test problems.

8.4.1 The Link-Biased Encoding

The idea of the *link-biased* LB encoding (Palmer, 1994; Rothlauf and Goldberg, 2003) is to represent a tree using a bias for each edge and to modify the edge weights w_{ij} according to these biases. A tree is constructed from the modified edge weights w'_{ij} by calculating an MST. Therefore, a genotype b of length $l = n(n-1)/2$ holds biases for the links. When constructing the phenotype (the tree) from the genotype, the biases are temporarily added to the original distance weights w_{ij}. To get the represented tree, an MST is calculated using the modified distance weights w'_{ij}. Links with low w'_{ij} will be used with high probability, whereas edges with high w'_{ij} will not appear in the tree. To finally calculate the tree's fitness, the encoded tree is evaluated by using the original edge weight matrix W and demand matrix R.

The weights b_{ij} are floating values between zero and one. The original distance weights w_{ij} are modified by the elements b_{ij} of the bias vector as

$$w'_{ij} = w_{ij} + P_1 b_{ij} w_{\max}, \tag{8.3}$$

where the w'_{ij} are the modified weights, $w_{\max}$ is the largest weight ($w_{\max} = \max(w_{ij})$), and P_1 is the *link-specific bias*. The parameter P_1 controls the influence of the biases b_{ij} and has a large impact on the structure of the encoded tree. For $P_1 = 0$, the b_{ij} have no influence and only an MST based on the w_{ij} can be represented. Prim's or Kruskal's algorithm can be used for generating an MST resulting in time that is $O(n^2)$ or $O(n^2 \log(n))$, respectively. The structure of the tree depends not only on the bias values b_{ij}, but also on the weights w_{ij}. Therefore, the same link-biased genotype can represent different trees if different weight matrices W are used.

We give an example of the construction of a tree from a bias vector b. For a tree with $n = 4$ nodes, the genotype has length $l = n(n-1)/2 = 6$. The link-biased genotype is $b = \{0.1, 0.6, 0.2, 0.1, 0.9, 0.3\}$. With $P_1 = 1$ and the edge weights $w = \{10, 30, 20, 40, 10, 20\}$, the modified weights are calculated according to (8.3) as

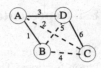

Fig. 8.8 An example tree.
The labels indicate the edge
numbers

$w' = \{14, 54, 28, 44, 46, 32\}$. Notice that $w_{max} = 40$. The represented tree that is calculated as the MST using the modified edge weights w' is shown in Fig. 8.8. The six possible edges are labeled from 1 to 6, and the tree consists of e_{AB} (link 1 with $w'_{AB} = 14$), e_{AD} (link 3 with $w'_{AD} = 28$), and e_{CD} (link 6 with $w'_{CD} = 32$).

The LB encoding is a specialized version of the more general *link and node biased* (LNB) encoding (Palmer, 1994). The functionality of the LNB encoding is analogous to the LB encoding. However, the LNB encoding uses additional biases for each node and the modified edge weights are calculated as

$$w'_{ij} = w_{ij} + P_1 b_{ij} w_{max} + P_2 (b_i + b_j) w_{max}, \tag{8.4}$$

with the elements b_i ($i \in [0, \ldots, n-1]$) of the node bias vector and the node-specific bias P_2. The tree T is calculated as an MST using the modified distance weights w'_i. In comparison to the LB encoding, the length of a genotype increases from $n(n-1)/2$ to $n(n+1)/2$.

Abuali et al (1995) compared the LNB encoding to some other representations for the probabilistic MST problem and in some cases found the best solutions by using the LNB encoding. Raidl and Julstrom (2000) proposed a variant of this encoding and observed solutions superior to those of several other representations for the degree-constrained MST problem. For the same type of problem, Krishnamoorthy and Ernst (2001) proposed yet another version of the LNB encoding. Gaube and Rothlauf (2001) found that for $P_2 > 0$, trees similar to stars are encoded with higher probability and for large P_2 not all possible trees can be encoded. For $P_2 \to \infty$ (and P_1 finite), only stars can be encoded. Therefore, the LNB encoding with additional node-biases and $P_2 \neq 0$ is only useful if optimal solutions are similar to stars.

In the following paragraphs, we focus on the LB encoding. The LB encoding is redundant as genotypes consisting of $n(n-1)/2$ real values b_{ij} encode a finite number of trees (phenotypes). For large values of the link-specific bias ($P_1 \to \infty$), the LB encoding becomes uniformly redundant (Sect. 6.2.2). This means every possible tree is represented by the same number of different genotypes. With decreasing P_1, the LB encoding becomes non-uniformly redundant and MST-like solutions are over-represented. Then, a random LB-encoded genotype represents MST-like trees with higher probability. If $P_1 = 0$, only one tree – an MST with respect to the w_{ij} – can be represented, since the elements b_{ij} of the bias vector have no impact on w'_{ij}.

Figure 8.9 illustrates how the LB encoding over-represents trees similar to an MST (Rothlauf and Goldberg, 2003). We plot the probability P_r that a link of a randomly generated genotype is part of an MST over the link-specific bias P_1. Results are presented for two tree sizes n (16 and 28 nodes) and for P_1 ranging from 0.01 to 1,000. We plot the mean and standard deviation of P_r. For large values of P_1, all n^{n-2} possible trees are uniformly represented. With decreasing P_1, randomly cre-

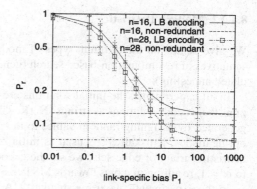

Fig. 8.9 Probability that a link of a randomly generated LB-encoded tree is part of an MST over the link-specific bias P_1

ated genotypes contain links that are also used in MSTs with higher probability. For small values of P_1, all edges of a randomly created individual are also with high probability P_r part of an MST. For $P_1 \rightarrow 0$, only an MST can be encoded.

If P_1 is too small, not all possible trees can be represented; How large must P_1 be to allow the LB encoding to represent all possible trees? Kruskal's algorithm, which is used to create the tree from the w'_{ij}, uses a sorted list of edges and iteratively inserts edges with lowest w'_{ij} into the tree. Because the LB encoding modifies the original edge weights w_{ij} (8.3), it also modifies this ordered list of edges and thus allows the encoding of trees different from MSTs. If $P_1 \geq 1$ (and all $b_{ij} = 1$), $P_1 b_{ij} \max(w_{ij})$ can be equal to or greater than the highest original edge weight $\max(w_{ij})$. Then, the encoding can completely change the order of the list of edges as for edge e_{lm} the original $w_{lm} = \min(w_{ij})$ can be changed to $w'_{lm} = \min(w_{ij}) + P_1 \max(w_{ij}) > \max(w_{ij})$ if $P_1 > 1$. As a result, it is possible to encode all n^{n-2} different trees using the LB encoding if $P_1 > 1$ (assuming a proper setting of b_{ij}).

Therefore, we make a recommendation for the choice of P_1 (Rothlauf, 2009c). With lower P_1, the LB encoding is biased towards MSTs. Thus, we expect higher performance for OCST problems. However, if $P_1 < 1$, some trees cannot be represented. Therefore, we recommend setting $P_1 \approx 1$. This ensures that all possible solutions can be encoded while still over-representing solutions similar to MSTs.

8.4.2 Experimental Results

This section presents experimental results for the LB encoding. We study the performance of several modern heuristics for OCST problems from the literature and randomly generated test instances. Furthermore, we examine how performance depends on P_1. In the experiments, we use the NetKey representation (Sect. 8.1.2) and the edge-set encoding (Sect. 8.3.1) as benchmark.

8.4.2.1 Modern Heuristics

We give details on the different types of modern heuristics that are used as representatives of recombination-based search (genetic algorithm) and local search (simulated annealing).

In all experiments, the initial solutions are generated randomly. For representations that encode solutions as strings (NetKey, LB encoding), the initial solutions are generated by assigning random values to the genotype strings. For edge-sets without heuristics, KruskalRST is used as initialization method (Sect. 8.3.1.1). For the heuristic variant of edge sets, we set the encoding-specific initialization parameter to $\alpha = 1$, resulting in a bias towards MST-like trees.

In the experiments, we use a standard GA (Sect. 5.2.1) with uniform crossover, no mutation, and standard 2-tournament selection without replacement as a representative for recombination-based search. A GA run is stopped after the population has fully converged (all solutions in the population encode the same phenotype), or after 100 generations. For string representations, uniform crossover chooses the alleles for the offspring randomly from the two parents. For edge-sets, we use the heuristic as well as non-heuristic crossover variant (Sect. 8.3.1).

In SA (Sect. 5.1.4), our local search operator randomly exchanges the position of two values in the encoded solution. For the LB encoding or NetKeys, two randomly chosen elements b_{ij} and b_{kl} of the bias vector are exchanged. Therefore, each search step assigns modified weights w' to two edges, resulting in a different encoding of a tree. In each iteration, the temperature T is reduced by a factor of 0.99 ($T_i = 0.99 T_{i-1}$). With lowering T, the probability of accepting worse solutions decreases. Each SA run is stopped if the number of search steps exceeds $iter$ or there are no improvements in the last $iter_{term}$ search steps. The initial temperature T_0 is set with respect to the target problem instance. For each instance, 1,000 unbiased trees T_i are generated and $T_0 = 2\sigma(w(T_i))$, where $\sigma(w(T_i))$ denotes the standard deviation.

8.4.2.2 GA Performance for Existing Problem Instances

Table 8.8 compares the GA performance using NetKeys, edge-sets with either naive or heuristic initialization and crossover operators, and the LB encoding with different link-specific biases P_1. We show results for problem instances from Sect. 8.1.3 and list success probabilities P_{suc} and average number of generations t_{conv} of a GA run. The GA is run 200 times for each of the test instances with population size N.

The results indicate that the GA performs well for the LB encoding with $P_1 \approx 1$. For some problem instances, a higher (palmer24) or lower (palmer6 and palmer12) value of P_1 can slightly improve GA performance, but choosing $P_1 \approx 1$ results on average in a good and robust GA performance. As expected, GA performance is low for small values of P_1 as only MSTs can be encoded (except for berry6 and berry35 where an MST is optimal (see Table 8.1, p.190)). For high values of P_1 ($P_1 = 100$), all possible trees are represented with the same probability, and GAs using the LB encoding show similar performance as with NetKeys. Then, no problem-

Table 8.8 Success probabilities P_{suc} and average number of generations t_{conv} of a GA using different types of representations

	N		Net Key	edge-sets		LB				
				naive	heur.	$P_1=0.05$	$P_1=0.5$	$P_1=1$	$P_1=2$	$P_1=100$
palmer6	16	P_{suc}	0.27	0.17	0	0.31	**0.84**	0.66	0.48	0.28
		t_{conv}	15.1	10.8	1.1	4.3	10.0	12.2	13.9	15.3
palmer12	300	P_{suc}	0.31	0.31	0	0	**0.74**	0.62	0.48	0.34
		t_{conv}	60.8	36.5	4.5	40.9	48.6	52.4	54.8	60.9
palmer24	800	P_{suc}	0.72	0.08	0	0	0.03	0.49	0.64	**0.77**
		t_{conv}	87.8	60.4	6.6	112.5	83.5	85.1	85.8	87.1
raidl10	70	P_{suc}	0.78	0.44	0	0	**1**	1	0.98	0.78
		t_{conv}	33.9	22.8	2.1	16.3	21.5	25.0	27.4	33.0
raidl20	450	P_{suc}	0.46	0.53	0	0	**0.99**	0.92	0.85	0.41
		t_{conv}	82.7	54.3	6.0	40.1	63.1	69.0	73.1	82.7
berry6	16	P_{suc}	0.54	0.38	1	1	**1**	0.98	0.82	0.56
		t_{conv}	14.2	10.4	1.5	1.0	6.9	9.4	11.6	14.5
berry35	300	P_{suc}	0.03	0	1	1	1	**1**	0.98	0.04
		t_{conv}	98.3	74.8	1.0	1.0	70.3	76.0	80.0	98.0

specific knowledge is exploited by the LB encoding. The benchmark representation edge-sets with heuristic search operators fails if optimal solutions are different from MSTs. When using the naive variant, the GA shows higher performance, but is on average outperformed by the NetKey encoding (Tzschoppe et al, 2004).

8.4.2.3 Influence of P_1

Figures 8.10(a) and 8.10(b) show the success probability P_{suc} of a GA and SA over the link-specific bias P_1 for 100 randomly created OCST problem instances with problem size $n = 16$ and randomly chosen edge weights $w_{ij} \in]0, 100]$. For each problem instance, 50 independent runs are performed and P_{suc} is averaged over all 100 problem instances. The GA has population size $N = 200$. The SA runs are stopped either after $iter = 20,000$ search steps, or there are no improvements in the last $iter_{term} = 2,000$ search steps.

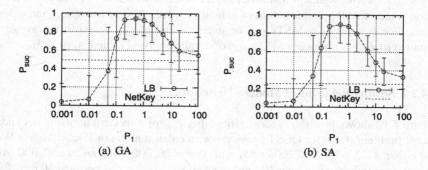

Fig. 8.10 Success probability P_{suc} of a GA and SA over the link-specific bias P_1

We find that GA and SA performance is best when using the LB encoding with $P_1 \approx 1$. A pairwise t-test is performed on the success probabilities P_{suc} of GA and SA; for $0.1 \leq P_1 \leq 10$, the LB encoding outperforms the unbiased NetKey encoding with high significance ($p < 0.001$). Furthermore, as expected, the performance of both modern heuristics is similar for the LB encoding with large P_1 and NetKeys. For large values of P_1, the LB encoding becomes uniformly redundant, all possible trees are represented with the same probability, and it is not possible to exploit problem-specific properties of OCST problems. For low values of P_1, P_{suc} becomes small as OCST problems can only be solved where the optimal solution is an MST.

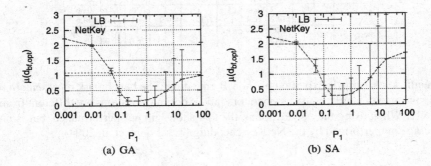

Fig. 8.11 Average distance $\mu(d_{bf,opt})$ between the best solutions T_{bf} that have been found by a GA and SA and the optimal solution T_{opt} over the link-specific bias P_1 for 100 random OCST problems

To better understand why the LB encoding shows maximum performance for $P_1 \approx 1$, Fig. 8.11 shows the average distance $\mu(d_{bf,opt})$ between the best solutions T_{bf} that have been found by the GA and the SA and the optimal solution T_{opt} for all 100 random 16 node OCST problems. The results confirm the previous findings. Because optimal solutions are similar to MSTs and the LB encoding encodes MST-like solutions with higher probability, $d_{bf,opt}$ decreases with lower P_1. However, if P_1 is too small, not all possible solutions can be encoded ($P_1 < 1$) and $d_{bf,opt}$ increases again. For $P_1 \rightarrow 0$ only MSTs can be encoded and $d_{bf,opt}$ becomes the average distance $\mu(d_{mst,opt})$ between MSTs and optimal solutions (Table 8.4, p. 193).

8.4.2.4 Scaling Behavior of Modern Heuristics

Figure 8.12 shows how the success probability P_{suc} of a modern heuristic depends on the problem size n for OCST problems with either random or Euclidean w_{ij}. We use either a GA with $N = 200$ or SA with $iter = 20,000$ and $iter_{term} = 2,000$. As before, we randomly generate 100 instances of each size n and perform 50 runs for each representation. We plot results for different values of P_1. The NetKey encoding is used as an example of a problem-independent (unbiased) encoding.

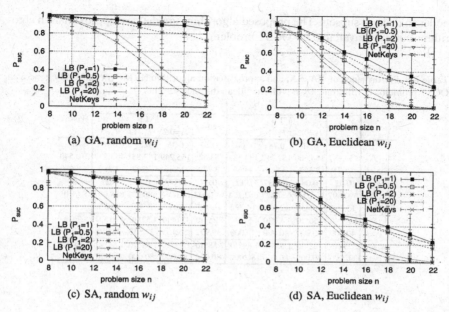

Fig. 8.12 Success probability P_{suc} of modern heuristics over n for random OCST problems

The plots show similar results for GA or SA as well as for OCST problems with random or Euclidean weights. As expected, the performance of representations that are biased towards MSTs is higher for OCST problems with random weights than with Euclidean weights (compare Fig. 8.2, p. 193). Then, GA and SA performance decreases with increasing n as the parameters of the modern heuristics are fixed and larger instances are more difficult to solve. To get better results, either larger population sizes (GA), or a different cooling schedule combined with a higher number of search steps (SA) would be necessary. More importantly, GA and SA performance is highest with $P_1 \approx 1$. Unbiased representations such as NetKeys are outperformed. Results for edge-sets are omitted due to the low performance of these encodings.

8.4.2.5 Performance of Modern Heuristics

Tables 8.9 and 8.10 present the performance of modern heuristics using the LB encoding for random OCST problems of different type and size and compare it to NetKeys and the results (denoted as PeRe) obtained by an approximation algorithm from Peleg and Reshef (1998) and Reshef (1999). They presented an $O(\log n)$ approximation for OCST problems with Euclidean distance weights and an $O(\log^3 n)$ approximation for arbitrary (non-Euclidean) distance weights. The approximation uses a recursive construction algorithm with a partitioning step that clusters the nodes in disjoint subsets. The proposed algorithms are the best approximation algorithms currently available for OCST problems.

Table 8.9 Performance of GA, SA, and approximation algorithm (PeRe) for randomly generated OCST problems of different size n with Euclidean distance weights

n		GA				SA				PeRe
		NetKey	LB $P_1=0.2$	$P_1=1$	$P_1=5$	NetKey	LB $P_1=0.2$	$P_1=1$	$P_1=5$	
10	P_{suc}	0.56	0.31	0.78	0.60	0.70	0.35	0.83	0.76	0
	$w(T_{bf})$	153,583	155,600	152,717	153,334	153,628	155,249	152,851	153,237	210,540
	eval	4,071	2,565	3,289	3,868	3,770	2,865	2,837	3,389	-
20	P_{suc}	0.02	0.06	0.28	0.12	0.04	0.06	0.33	0.21	0
	$w(T_{bf})$	690,697	692,340	677,477	682,193	691,633	693,085	680,821	684,057	$1.1*10^6$
	eval	15,207	10,438	12,690	14,292	16,496	7,647	8,350	12,776	-
50	$w(T_{bf})$	$4.78*10^6$	$4.56*10^6$	$4.47*10^6$	$4.51*10^6$	$5.25*10^6$	$4.56*10^6$	$4.52*10^6$	$4.61*10^6$	$9.7*10^6$
	eval	40,000	40,000	40,000	40,000	39,992	35,103	37,538	39,995	0
100	$w(T_{bf})$	$2.21*10^7$	$1.85*10^7$	$1.83*10^7$	$1.89*10^7$	$2.58*10^7$	$1.83*10^7$	$1.82*10^7$	$1.88*10^7$	$5.34*10^7$
	eval	80,000	80,000	80,000	80,000	80,000	80,000	80,000	80,000	-

Table 8.10 Performance of GA, SA and approximation algorithm (PeRe) for randomly generated OCST problems of different size n with random distance weights

n		GA				SA				PeRe
		NetKey	LB $P_1=0.2$	$P_1=1$	$P_1=5$	NetKey	LB $P_1=0.2$	$P_1=1$	$P_1=5$	
10	P_{suc}	0.84	0.86	0.97	0.90	0.91	0.88	0.96	0.93	0
	$w(T_{bf})$	6,824	6,799	6,789	6,800	6,812	6,798	6,789	6,800	19,438
	eval	3,665	1,867	2,722	3,355	3,652	2,293	2,599	3,124	-
20	P_{suc}	0.11	0.94	0.89	0.57	0.14	0.86	0.84	0.61	0
	$w(T_{bf})$	19,946	18,738	18,746	18,908	20,059	18,803	18,794	19,017	88,069
	eval	14,113	8,215	10,809	12,626	16,645	5,547	7,384	11,721	-
50	$w(T_{bf})$	108,465	65,256	65,610	69,213	196,147	65,394	66,405	78,640	592,808
	eval	40,000	36,612	40,000	40,000	39,990	25,134	37,089	39,903	-
100	$w(T_{bf})$	$7.7*10^5$	$1.6*10^5$	$1.8*10^5$	$2.3*10^5$	$1.4*10^6$	$1.7*10^5$	$2.0*10^5$	$3.1*10^5$	$2.4*10^6$
	eval	80,000	80,000	80,000	80,000	80,000	79,862	80,000	80,000	-

The table shows the average cost $w(T_{bf})$ of the best solution found, and the average number of fitness evaluations *iter* for randomly created problems with 10, 20, 50, and 100 nodes. For each n, we generate 100 random problem instances and perform 20 runs for each instance (for $n = 100$, we only create 15 instances due to the high computational effort). The distance weights w_{ij} are either random or Euclidean. The demands r_{ij} are generated randomly and are uniformly distributed in $]0, 10]$. For small problem instances ($n = 10$ and $n = 20$) optimal solutions are determined using the iterated GA described in Sect. 8.1.3.2.

Because the performance of modern heuristics depends on parameter settings and modern heuristics with fixed parameters show lower performance with increasing problem size n (compare Fig. 8.12), the GA and SA parameters have to be adapted with increasing n. Table 8.11 lists how the GA (population size N) and SA (maximal number of iterations $iter$, and termination criterion $iter_{term}$) parameters are chosen with respect to n. By using higher population sizes (GA), or a larger number of search steps (SA), solutions with higher quality can be found for larger n.

Table 8.11 GA and SA parameters

	n	10	20	50	100
GA	N	100	200	400	800
SA	$iter$	10,000	20,000	40,000	80,000
	$iter_{term}$	2,000	4,000	8,000	16,000

We find that a GA and SA using the LB encoding with $P_1 \approx 1$ outperform the unbiased NetKey encoding and the approximation algorithm of Peleg and Reshef (PeRe). The approximation algorithm never finds optimal or near-optimal solutions and the solution quality is low. A GA and SA using NetKeys show good performance for smaller instances, but with increasing n solutions with lower quality are found and the performance differences between the problem-specific LB encoding and the problem-independent NetKeys encoding become greater. A comparison between the results for Euclidean and random w_{ij} shows that for Euclidean w_{ij}, GA and SA performance is maximal for $P_1 = 1$. Then, MSTs are over-represented and high quality solutions can be found which are similar to an MST. For OCST problems with random w_{ij}, SA and GA performance can be slightly improved by lowering P_1 to 0.2. The performance improvement is due to the fact that for random w_{ij}, optimal solutions are more MST-like than for Euclidean w_{ij} (compare Fig. 8.2, p. 193). Therefore, modern heuristics that use representations with a stronger bias towards MSTs are expected to show higher performance than unbiased representations, and a GA and SA with $P_1 = 0.2$ on average show higher performance than $P_1 = 1$.

We see that OCST problems can be solved effectively using the redundant LB encoding with $P_1 \approx 1$. This setting of P_1 results in robust behavior of modern heuristics and allows them to appropriately exploit the bias of OCST problems. The high performance of the LB encoding is due to the over-representation of MST-like solutions and can be observed for mutation-based as well as recombination-based search.

8.5 Initialization

This section studies how the choice of a biased initial solution affects the performance of modern heuristics. We have seen in Sect. 8.1.3 that optimal solutions of OCST problems are similar to an MST. This similarity suggests that the performance of modern heuristics can be improved by starting with an MST.

Consequently, we investigate how the performance of different modern heuristics depends on the type of initial solution. We show results for a greedy search strategy, local search, SA, and a GA. Greedy search and local search are representative examples of improvement heuristics without any diversification elements. SA and GAs are representative examples of modern heuristics that are based on either local search or recombination-based search. We find that the performance of modern heuristics can be improved by starting with an MST (Rothlauf, 2009a).

Section 8.5.1 discusses other optimization problems where starting from an MST increases the performance of modern heuristics. Sect. 8.5.2 presents experimental results indicating that the performance of modern heuristics for the OCST problem can be improved if starting with an MST.

8.5.1 Using an MST as Initial Solution

The idea to exploit problem-specific knowledge about an optimization problem by starting search from an MST has also been used for the design of modern heuristics for other tree and tree-related problems. For example, for the metric TSP, creating an Eulerian graph from the MST (each edge is taken twice) and producing an Eulerian tour on this graph results in a tour whose cost is at most twice the cost of the optimal tour (compare Vazirani (2003, p. 31f)). Also, the 3/2-approximation algorithm of Christofides (1976) (Sect. 3.4.2) follows this concept. It starts with an MST, computes a minimum cost perfect matching on the set of odd-degree vertices, and then generates an Eulerian graph.

For the Steiner tree problem, using an MST as the starting solution is a promising strategy for modern heuristics as constructing an MST on the set of required nodes already results in a 2-approximation (Choukhmane, 1978). Therefore, different modern heuristics (e.g. the ER heuristic of Borah et al (1999)) start with an MST. Robins and Zelikovsky (2005) presented an efficient polynomial-time approximation scheme for the Steiner tree problem that starts with an MST and results in the best-known performance ratio approaching $1 + \ln 3/2 \approx 1.55$. Hwang (1976) showed for the rectilinear Steiner tree problem that the MST is a 3/2-approximation. Consequently, a variety of approaches either start with an MST and improve this solution in subsequent steps (Ho et al, 1990; Chao and Hsu, 1994; Julstrom, 2002; Robins and Zelikovsky, 2005), or imitate the MST construction of Kruskal and Prim (Bern, 1988; Richards, 1989). Arora (1998) found that Euclidean and rectilinear minimum-cost Steiner trees (as well as Euclidean TSPs) can be efficiently approximated arbitrarily close to optimality. For more information on heuristics for Steiner tree problems, we refer to Kahng and Robins (1992) and Robins and Zelikovsky (2005).

Another example where starting search with an MST results in efficient modern heuristics is the *capacitated MST problem*. The objective is to find a minimum cost tree such that some capacity constraints (usually on node weights) are observed. As a feasible (capacity constraints are met) MST is optimal, many simple heuristics consider edge weights and either construct a solution similar to the MST (e.g. the unified algorithm of Kershenbaum and Chou (1974), or the modified Kruskal algorithm of Boorstyn and Frank (1977)), or modify an infeasible MST solution (e.g. the start procedure of Elias and Ferguson (1974)). Examples of modern heuristics for the capacitated MST problem that start with an MST are presented in Kershenbaum et al (1980), and Gavish et al (1992). For a survey on heuristics and modern heuristics for the capacitated MST, we refer to Amberg et al (1996) and Voß (2001).

8.5.2 Experimental Results

We study how the choice of the initial solution affects the performance of various modern heuristics.

8.5.2.1 Search Strategies

In the experiments, we use greedy search (GS), simulated annealing (SA), local search (LS), and genetic algorithms (GA). Greedy search (Sect. 3.3.2.3) starts with an initial solution T_s and iteratively examines all neighboring trees that are different in one edge and chooses the tree T_j with lowest cost $w(T_j)$. The neighborhood of T_i consists of all T_j that can be created by the change of one edge ($d_{ij} = 1$). GS stops if the current solution cannot be improved any more. The number of different neighbors of a tree depends on its structure and varies between $(n-1)(n-2)$ and $\frac{1}{6}n(n-1)(n+1) - n + 1$ (Rothlauf, 2006).

In our SA (Sect. 3.4.3), in each iteration i, the temperature is reduced ($T_i = 0.995 \times T_{i-1}$). The initial temperature T_0 is set with respect to the different test problems to be solved. For each test problem, 1,000 random solutions T_i are generated before the SA run and $T_0 = 0.1 \times \sigma(w(T_i))$, where $\sigma(w(T_i))$ denotes the standard deviation of the cost of T_i. T_0 is set relatively low as large values of T_0 would result in a high probability of accepting worse solutions at the beginning of an SA run and thus search performance would become independent of T_s. Each SA run is stopped after *iter* search steps.

In local search, a new, neighboring solution replaces the original solution if it has lower cost. This approach is equivalent to SA with $T = 0$. Like the other heuristics, LS is stopped after *iter* search steps.

The conventional steady-state GA (Sect. 5.2.1) uses edge-sets (Sect. 8.3.1) with and without heuristics and is designed as described in Sect. 8.3.3.1. The non-heuristic variant of edge-sets (denoted GA) uses non-heuristic search operators (Sect. 8.3.1.1, p. 196). The heuristic variant (denoted GA-ESH) uses heuristic initialization ($\alpha = 1.5$), mutation, and crossover operators (Sect. 8.3.1.2, p. 197). For the experiments, we use a population size $N = 50$, a mutation probability of $p_{mut} = 1/n$, and stop the GA after *iter* evaluations.

Peleg and Reshef (1998) and Reshef (1999) presented an $O(\log n)$ approximation for OCST problems with Euclidean distance weights and an $O(\log^3 n)$ approximation for arbitrary (non-Euclidean) distance weights. The proposed algorithms are the best approximation algorithms that are currently available for the OCST problem.

In the experimental study, we compare the performance of different modern heuristics for three different starting solutions T_s: either the solution generated by the approximation algorithm of Peleg and Reshef (1998) (denoted PeRe), or a random tree (denoted rnd), or an MST.

8.5.2.2 Performance for Existing Test Problems

Table 8.12 presents results for the test instances from Sect. 8.1.3. It lists for different starting solutions T_s, the cost $w(T)$ of 1,000 randomly generated solutions of type T_s (denoted random), the performance (average cost of best found solution and average number of evaluations) of a greedy search starting with T_s, the maximum number *iter* of evaluations for SA, LS, GA, and GA-ESH, and the average cost of the best found solution for SA, LS, GA, and GA-ESH. T_s is either an MST, a random tree, or the result of the PeRe-approximation. For GA-ESH, we only present results for heuristic initialization (Sect. 8.3.1.2, p. 197) which creates MST-like solutions. For the population-based GA, one randomly chosen individual in the initial population of size N is set to T_s. Therefore, this randomly chosen solution is either an MST, a randomly chosen tree, or the result of the PeRe-approximation. All other solutions in the initial population are random trees. 50 independent runs are performed for each problem instance and the gap $(w(T^{bf}) - w(T^{best}))/w(T^{best})$ (in %) between the best found solution T^{bf} and T^{best} (Table 8.1) is shown in brackets.

For all problem instances, the average cost of randomly created solutions and PeRe is greater than the cost of an MST. For the greedy heuristic, there are no large differences between the cost of the best found solution for different T_s. However, there are large differences in the average number *eval* of evaluations. Starting from a random tree always results in highest *eval*; starting from an MST always results (except for palmer24) in lowest *eval*. On average, *eval* for starting from PeRe is between the results for MST and random tree. For SA, LS, and GA, starting with an MST always results (except for palmer24 and raidl10) in better solutions than starting from a random tree, and often in better solutions than starting from PeRe. Using SA or LS and starting from an MST always results (except palmer6 and berry35) in the same (berry6) or better performance than GA-ESH. Summarizing the results, the cost of an MST is on average lower than the cost of the PeRe-approximation (up to a factor of six for raidl100) and the cost of a random tree (up to a factor of 35 for raidl100). The modern heuristics find better solutions (up to a factor of four for raidl75) when starting with an MST in comparison to starting with a random tree.

8.5.2.3 Performance for Random OCST Problems Problems

Tables 8.13 and 8.14 extend the analysis and show results for random OCST problems with 10, 25, 50, 100, and 150 nodes. We report the same performance figures as in the above paragraphs (Table 8.13). Table 8.15 lists the maximum number of evaluations allowed for SA, LS, GA, and GA-ESH.

Table 8.12 Performance of different modern heuristics using different starting solutions T_s for test instances from the literature

	T_s	random cost	greedy cost (%opt)	eval	iter	SA cost (%opt)	LS cost (% opt)	GA cost (%opt)	GA-ESH cost (%opt)
palmer6	MST	709,770	**693,180** (0)	57		698,644 (0.8)	699,740 (0.9)	693,680 (0.07)	**693,180** (0)
	rnd	$1{,}7*10^6$	699,442 (0.9)	186	500	700,150 (1.0)	700,635 (1.1)	693,781 (0.09)	
	PeRe	926,872	695,566 (0.3)	75		700,927 (1.1)	697,951 (0.7)	693,478 (0.04)	
palmer12	MST	$3{,}8*10^6$	3,541,915 (3.3)	1,794		3,579,384 (4.4)	3,498,458 (2.0)	3,730,657 (8.8)	3,623,253 (5.7)
	rnd	$1{,}1*10^7$	**3,492,357** (1.9)	5,752	500	3,594,002 (4.8)	3,512,364 (2.5)	4,004,888 (17)	
	PeRe	$4{,}6*10^6$	3,522,282 (2.7)	2,806		3,580,330 (4.4)	3,502,914 (2.2)	3,912,014 (14)	
palmer24	MST	$1{,}9*10^6$	**1,086,656** (0)	57,654		1,098,379 (1.1)	1,092,353 (0.5)	1,154,449 (6.2)	1,250,073 (15)
	rnd	$1{,}0*10^7$	**1,086,656** (0)	92,549	2,500	1,097,179 (1.0)	1,098,732 (1.1)	1,226,783 (13)	
	PeRe	$2{,}3*10^6$	**1,086,656** (0)	39,303		1,103,834 (1.6)	1,096,615 (0.9)	1,181,316 (8.7)	
raidl10	MST	58,352	**53,674** (0)	435		54,762 (2.0)	53,699 (0.05)	57,141 (6.5)	55,761 (3.9)
	rnd	328,993	**53,674** (0)	2,296	500	54,663 (1.8)	54,009 (0.6)	67,893 (26)	
	PeRe	194,097	**53,674** (0)	1,711		54,796 (2.1)	**53,674** (0)	70,120 (31)	
raidl20	MST	168,022	**157,570** (0)	3,530		158,983 (0.9)	**157,570** (0)	159,911 (1.5)	158,974 (0.9)
	rnd	$1{,}9*10^6$	157,995 (0.3)	50,843	2,500	161,023 (2.2)	160,214 (1.7)	205,718 (30.6)	
	PeRe	849,796	158,704 (0.7)	32,053		160,943 (2.1)	160,578 (1.9)	209,731 (33)	
raidl50	MST	912,303	809,311 (0.30)	211,394		829,780 (2.84)	811,098 (0.52)	852,091 (5.6)	880,927 (9.18)
	rnd	$2{,}0*10^7$	**806,946** (0.01)	$1{,}7*10^6$	10,000	864,736 (7.17)	887,066 (9.94)	1,541,047 (91)	
	PeRe	$5{,}9*10^6$	807,353 (0.06)	$1{,}0*10^6$		890,082 (10.3)	883,483 (9.50)	1,488,774 (85)	
raidl75	MST	$2{,}4*10^7$	**1,717,491** (0)	$1{,}3*10^6$		2,042,603 (19)	1,852,905 (7.9)	1,971,638 (15)	2,003,433 (17)
	rnd	$5{,}8*10^7$	**1,717,491** (0)	$8{,}4*10^6$	10,000	2,401,151 (40)	2,330,226 (36)	8,814,074 (413)	
	PeRe	$1{,}3*10^7$	1,749,322 (1.9)	$4{,}5*10^6$		2,370,276 (38)	2,303,405 (34)	4,957,022 (189)	
raidl100	MST	$3{,}6*10^6$	**2,561,543** (0)	$2{,}4*10^6$		2,713,040 (6)	2,619,256 (2.3)	2,831,167 (11)	2,935,381 (14.6)
	rnd	$1{,}1*10^8$	2,603,146 (1.6)	$2{,}2*10^7$	40,000	2,870,197 (12)	2,937,911 (15)	5,200,334 (103)	
	PeRe	$2{,}4*10^7$	2,709,603 (5.8)	$1{,}2*10^7$		2,941,064 (15)	2,959,953 (16)	4,620,145 (80)	
berry6	MST	**534**	**534** (0)	28		**534** (0)	**534** (0)	**534** (0)	**534** (0)
	rnd	1,284	**534** (0)	207	500	**534** (0)	**534** (0)	534.04 (0.07)	
	PeRe	842	**534** (0)	120		**534** (0)	**534** (0)	536 (0.3)	
berry35	MST	**16,915**	**16,915** (0)	3,387		21,818 (29)	**16,915** (0)	**16,915** (0)	**16,915** (0)
	rnd	379,469	**16,915** (0)	467,008	2,500	22,426 (33)	22,642 (34)	44,661 (164)	
	PeRe	60,382	**16,915** (0)	171,425		21,563 (27)	20,777 (23)	31,765 (88)	

Table 8.13 Performance of different modern heuristics using different starting solutions T_s for randomly created OCST test instances with Euclidean distance weights

n	T_s	random cost	greedy cost (%opt)	eval	SA cost (%opt)	LS cost (% opt)	GA cost (%opt)	GA-ESH cost (%opt)
10	MST	1,670	**1,515** (0%)	698	1,527 (0.80%)	1,539 (1.58%)	1,605 (5.94%)	1,574 (3.89%)
	rnd	3,284	1,520 (0.32%)	2,527	1,555 (2.65%)	1,550 (2.28%)	1,673 (10.4%)	
	PeRe	2,114	1,519 (0.23%)	1,475	1,530 (1.00%)	1,549 (2.20%)	1,665 (9.89%)	
25	MST	10,794	**8,839** (0%)	41,854	8,937 (1.11%)	8,885 (0.52%)	9,230 (4.43%)	9,412 (6.49%)
	rnd	40,261	8,844 (0.06%)	122,463	9,004 (1.88%)	9,027 (2.14%)	10,337 (17.0%)	
	PeRe	12,827	8,855 (0.18%)	62,073	9,021 (2.06%)	9,044 (2.33%)	11,088 (25.5%)	
50	MST	61,450	**44,142** (0%)	786,904	44,749 (1.38%)	44,600 (1.04%)	45,320 (2.67%)	47,391 (7.36%)
	rnd	207,139	73,179 (65.8%)	999,800	45,187 (2.37%)	45,364 (2.77%)	50,336 (14.0%)	
	PeRe	63,638	45,301 (2.63%)	778,602	45,248 (2.51%)	45,312 (2.65%)	51,592 (16.9%)	
100	MST	272,987	256,149 (42%)	10^7	181,538 (0.5%)	**180,592** (0%)	182,787 (1.2%)	189,559 (4.97%)
	rnd	1,207,970	1,069,412 (492%)	10^7	183,188 (1.4%)	182,706 (1.2%)	194,423 (7.7%)	
	PeRe	251,198	255,158 (41%)	10^7	182,419 (1.0%)	183,280 (1.5%)	197,270 (9.2%)	
150	MST	645,042	624,139 (56%)	10^7	399,312 (0.0%)	**399,262** (0%)	401,127 (0.5%)	417,045 (4.45%)
	rnd	3,390,500	3,082,983 (672%)	10^7	403,119 (1.0%)	402,687 (0.9%)	425,814 (6.7%)	
	PeRe	562,282	563,160 (41%)	10^7	401,615 (0.6%)	401,946 (0.7%)	434,682 (8.9%)	

Table 8.14 Performance of different modern heuristics using different starting solutions T_s for randomly created OCST test instances with uniformly distributed random distance weights

n	T_s	random cost	greedy cost (%opt)	eval	SA cost (%opt)	LS cost (% opt)	GA cost (%opt)	GA-ESH cost (%opt)
	MST	717	**676** (0%)	291	686 (1.37%)	680 (0.49%)	703 (3.93%)	689
10	rnd	2,863	677 (0.10%)	2,472	691 (2.11%)	685 (1.25%)	852 (25.9%)	(1.95%)
	PeRe	1,949	677 (0.11%)	1,661	687 (1.62%)	684 (1.10%)	855 (26.4%)	
	MST	3,094	**2,720** (0%)	12,420	2,806 (3.16%)	2,738 (0.65%)	2,840 (4.40%)	2,855
25	rnd	37,254	2,723 (0.08%)	126,859	2,893 (6.34%)	2,882 (5.93%)	4,894 (79.9%)	(4.93%)
	PeRe	14,000	2,723 (0.09%)	74,831	2,885 (6.03%)	2,864 (5.29%)	5,290 (94.5%)	
	MST	7,921	**6,663** (0%)	169,398	6,989 (4.89%)	6,753 (1.36%)	7,081 (6.28%)	7,205
50	rnd	214,238	45,822 (587%)	10^7	7,447 (11.8%)	7,435 (11.6%)	16,009 (140%)	(8.14%)
	PeRe	59,304	8,968 (34.6%)	949,768	7,492 (12.4%)	7,405 (11.1%)	14,523 (118%)	
	MST	21,719	18,436 (9.22%)	10^7	17,584 (4.2%)	**16,879** (0%)	18,020 (6.8%)	18,561
100	rnd	1,206,920	990,391 (5,767%)	10^7	18,882 (11%)	19,765 (17%)	42,324 (151%)	(10.0%)
	PeRe	244,349	187,958 (1,014%)	10^7	19,003 (13%)	19,432 (15%)	36,938 (119%)	
	MST	40,596	39,184 (29%)	10^7	31,451 (4.0%)	**30,247** (0%)	32,807 (8.5%)	33,350
150	rnd	3,290,630	3,094,862 (10,131%)	10^7	36,107 (19%)	36,712 (21%)	84,711 (180%)	(10.3%)
	PeRe	549,068	505,367 (1,570%)	10^7	37,417 (24%)	37,355 (24%)	71,298 (136%)	

Table 8.15 Maximum number of evaluations

n	10	25	50	100	150
$iter$	500	2,500	10,000	40,000	80,000

For Table 8.13, we generate random problem instances with Euclidean weights. The nodes are placed randomly on a two-dimensional grid of size 10×10. For Table 8.14, the weights are randomly generated and uniformly distributed in $]0,10]$. For both tables, the demands r_{ij} are randomly generated and uniformly distributed in $]0,10]$. The results for the randomly created OCST problem instances are similar to the results for the problem instances from the literature. The average cost of an MST is always lower than the average cost of random solutions (significant using a t-test with an error level of 0.01). Because for Euclidean weights (Table 8.13) the PeRe approximation is tighter and because optimal solutions have a larger average distance $\mu(d_{mst,opt})$ to an MST (compare Fig. 8.2), the PeRe approximation results in better solutions and the results are similar to an MST. For random weights (Table 8.14), the average cost of an MST is always lower than the average cost of the PeRe approximation (also significant using a t-test with an error level of 0.01).

For Euclidean weights (Table 8.13), greedy search starting from an MST either finds good solutions faster ($n < 100$), or finds better solutions ($n \geq 100$) than starting from a random tree. With increasing n, greedy search starting from PeRe performs similarly to when starting from an MST. For SA, LS, and GA, starting from an MST always finds better solutions than starting from a random solution or from PeRe. Furthermore, starting from an MST results in better results than GA-ESH (except $n = 10$).

For random weights (Table 8.14), starting from an MST always results in better performance than starting from a random tree or PeRe (significant using a t-test with an error level of 0.01, except SA with $n = 10$). Furthermore, using SA, LS, or GA and starting with an MST results in better performance than GA-ESH (except $n = 10$).

Chapter 9
Summary

This textbook taught us the art of systematically designing efficient and effective modern heuristics. We learned when to use modern heuristics, how to deliberately choose among the available types of modern heuristics, what are the relevant design elements and principles of modern heuristics, and how we can improve their performance by considering problem-specific knowledge. We want to summarize the main lessons learned.

For which types of problems should we use modern heuristics? There are not only modern heuristics, but we can usually choose from a variety of exact and heuristic optimization methods (see Chap. 3). After we have obtained a model of our problem, we have to select a proper optimization method. For problems that can be solved with polynomial effort (e.g. continuous linear problems, Sect. 3.2), modern heuristics are usually not appropriate. If the problem at hand is more difficult (e.g. NP-hard, Sect. 2.4.1), we recommend searching the literature for fast exact optimization methods (Sect. 3.3), fast heuristics (Sect. 3.4.1), or efficient approximation methods (Sect. 3.4.2). If such methods are available, we should try them; often modern heuristics are not necessary. For example, the knapsack problem is FPTAS-hard (Fig. 3.20, p. 90) and, thus, efficient solution approaches are available.

The situation becomes different if our problem is difficult and/or non-standard. Examples of difficult problems are APX-hard problems (Sect. 2.4.1) like the MAX SAT problem, the symmetric TSP, or the OCST problem. For such problems, modern heuristics are often a good choice, especially if problems become large. Furthermore, modern heuristics are often the method of choice for non-standard real-world problems which are different from the well-defined problems that we can find in the literature (Sect. 4.1). Real-world problems often have additional constraints, additional decision variables, and other optimization goals. We can try to reduce complexity and make our problem more standard-like by removing constraints, simplifying the objective function, or limiting the number of decision variables. However, we often do not want to neglect some important aspects and are not happy with the simplified model. In this case, modern heuristics are often the right choice.

How can we select a modern heuristic that fits our problem well? The locality of a problem (Sect. 2.4.2) describes how well distances between solutions correspond to their fitness differences. Locality has a strong impact on the performance of local search methods. High locality allows local search to find high-quality solutions in the neighborhood of already found good solutions and guide local search methods towards optimal solutions. In contrast, if a problem has low locality, local search methods cannot make use of information gathered in previous search steps but behave like random search. The decomposability of a problem (Sect. 2.4.3) describes how well a problem can be decomposed into smaller and independent sub-problems. The decomposability of a problem is high if the structure of the objective function is such that there are groups of decision variables that can be set independently of decision variables contained in other groups. It is low if it is not possible to decompose a problem into sub-problems that have little interdependence. Problems with high decomposability can be solved well using recombination-based modern heuristics because solving a number of smaller sub-problems is usually easier than solving the larger, original problem. Decomposable problems usually have high locality. For decomposable problems, the change of a decision variable changes the fitness of only one sub-problem. If the fitness of the other sub-problems remains unchanged, similar solutions often have similar fitness and, thus, the locality of the problem is high.

As many real-world problems have high locality and are decomposable, local as well as recombination-based search often show good performance and return high-quality solutions. However, direct comparisons between local and recombination-based search are only meaningful for particular problem instances and general statements on the superiority of one of these search concepts are unjustified. Which one is more appropriate for solving a particular problem instance depends on the specific characteristics of the problem (locality versus decomposability).

What are common design elements of modern heuristics? In Chap. 4, we discussed the design elements representation, search operator, fitness function, and initial solution. These elements are relevant for all different types of modern heuristics. Chapter 5 studied the fifth design element, which is the search strategy. The search strategy defines the intensification and diversification mechanisms.

Given a set of solutions, we can define a search space either by defining search operators or by selecting a metric (Sect. 2.3.1). Local search operators usually create neighboring solutions. Recombination operators generate offspring where the distances between offspring and parents are usually equal to or smaller than the distances between parents. Therefore, defining search operators implies a neighborhood relationship for the search space, and vice versa. There are two complementary approaches for designing representations and search operators: We may use (indirect) representations where solutions are encoded in a standard data structure (Sect. 4.2.4), and standard operators (Sect. 4.3.5) are applied to these genotypes. Then, a proper choice of the genotype-phenotype mapping (Sect. 4.2.3) is important for the performance of search. In contrast, a direct representation (Sect. 4.3.4) encodes solutions in their most natural problem space and designs search operators to operate on this search space. Then, no mapping between genotypes and phenotypes needs to be specified, but the search operators are specially designed for the phenotypes and are problem-specific (Sect. 4.3.1).

Designing a fitness function (Sect. 4.4) and initialization method (Sect. 4.5) is usually easier than designing proper representations and search operators. The fitness function is determined by the objective function and must allow modern heuristics to perform pairwise comparisons between solutions. Initial solutions are usually randomly created if no a priori knowledge about the problem exists.

Finally, search strategies define intensification and diversification mechanisms used during search. Search strategies must ensure that the search stays focused (intensification) but also allow the search to escape from local optima (diversification). This is achieved by various diversification techniques based on the representation, search operator, fitness function, initialization, or explicit diversification steps controlled by the search strategy (Chap. 5).

How can we categorize different types of modern heuristics? Modern heuristics are extended variants of improvement heuristics (Sect. 3.4.3). Both consider information about previously sampled solutions for future search decisions. However, in contrast to improvement heuristics which only perform intensification steps, modern heuristics also allow inferior solutions to be generated during search. Therefore, modern heuristics use intensification as well as diversification steps during search. As the behavior of modern heuristics is defined independently of the problem, they are general-purpose methods applicable to a wide range of problems.

Diversification allows search to escape from local optima whereas intensification ensures that the search moves in the direction of solutions with higher quality. Diversification is often the result of large modifications of solutions. Intensification uses the fitness of solutions to guide search and ensures that the search moves in the direction of solutions with higher fitness. Existing local and recombination-based search strategies mainly differ in the way in which they control diversification and intensification (Sect. 5.1). Based on the design elements, there are different strategies to introduce diversification into the search and to escape from local optima:

- By using different types of neighborhoods during search, it is possible to escape from local optima and explore larger areas of the search space. Different neighborhoods can be the result of changing genotype-phenotype mappings or search operators during search (Sect. 5.1.1). Examples are variable neighborhood search, problem space search, the rollout algorithm, or the pilot method.
- Modifications of the fitness function can also lead to diversification (Sect. 5.1.2). An example is guided local search which systematically changes the fitness function with respect to the progress of search.
- Diversity can be introduced by performing repeated runs with different initial solutions (Sect. 5.1.3). Examples are iterated descent, large-step Markov chains, iterated Lin-Kernighan, chained local optimization, or iterated local search.
- Finally, the search strategy explicitly controls diversification and intensification (Sect. 5.1.4). Examples of search strategies that use a controlled number of large search steps towards solutions of lower quality to increase diversity are simulated annealing, threshold accepting, or stochastic local search. Representative examples of strategies that consider previous search steps for diversification are tabu search or adaptive memory programming.

We can apply the same categorization also to recombination-based search methods and distinguish methods with respect to the different intensification and diversification mechanisms used.

What are the basic principles for the design of modern heuristics? A fundamental assumption about the application of modern heuristics is that the vast majority of real-world problems are neither deceptive nor difficult (Sect. 2.4.2.1) and have high locality (Sect. 2.4.2). We assume that deceptive problems have no importance in the real world as usually optimal solutions are not isolated in the search space surrounded by only low-quality solutions. Furthermore, we assume that the metric of a search space is meaningful and, on average, the fitness differences between neighboring solutions are smaller than between randomly chosen solutions. Only because most real-world problems are neither difficult nor deceptive, guided search methods that use information about previously sampled solutions can outperform random search. This assumption is reasonable as the discussion about the no-free-lunch theorem (Sect. 3.4.4) shows that modern heuristics perform like random search on problems that are difficult or have a trivial topology, and even worse on deceptive problems.

Therefore, search operators and representations must fit the metric of the search space. If local as well as recombination-based search operators are not able to generate similar solutions, intensification of search is not possible and modern heuristics behave like random search. For a representation which introduces an additional genotype-phenotype mapping, we must make sure that it does not alter the character of the search operators. Therefore, its locality must be high; this means the phenotype metric must fit the genotype metric. Low locality of a representation randomizes the search and leads to low performance of modern heuristics.

In Chap. 7, we studied the importance of high-locality representations for grammatical evolution versus standard genetic programming. Grammatical evolution is a variant of genetic programming using as genotypes linear strings instead of parse trees. The case study illustrates that the locality of the representation/operator combination used in grammatical evolution is lower than for standard genetic programming. Consequently, the performance of local search is reduced.

How can we use problem-specific knowledge for the design of modern heuristics? There is a general trade-off between the effectiveness and application range of optimization methods. Usually, the more problems that can be solved with one particular optimization method, the lower its resulting average performance. Therefore, standard modern heuristics that are not problem-specific often only work for small or toy problems. As the problem gets larger and more realistic, performance degrades. To improve the performance for selected optimization problems, we must design modern heuristics in a more problem-specific way.

By assuming that most problems in the real world have high locality, modern heuristics already exploit a specific property of problems, namely their high locality. Only by assuming that problems have high locality, modern heuristics can outperform random search (see the discussion about the no-free-lunch theorem and subthreshold-seeking behavior from Sect. 3.4.4, p. 101). We can further increase the performance of modern heuristics if we have some idea about properties of good or even bad solutions to our problem. Such problem-specific knowledge can be exploited by introducing a bias into modern heuristics. The bias should consider this knowledge and, for example, concentrate search on solutions that are expected to be of high quality or avoid solutions expected to be of low quality. A bias can be considered in all design elements of modern heuristics, namely the representation, the search operator, the fitness function, the initialization, and also the search strategy. However, we recommend biasing modern heuristics only if we have obtained some particular knowledge about an optimization problem or problem instance. If we have no knowledge about properties of a problem, we should not bias modern heuristics as this would mislead search heuristics.

Chapter 8 presented a case study on the design of problem-specific modern heuristics for the optimal communication spanning tree (OCST) problem. We find that optimal solutions for OCST problems are similar to optimal solutions for the (much simpler) minimum spanning tree problem. Thus, biasing the representation, search operator, initial solution, or fitness function towards the minimum spanning tree can increase the performance of modern heuristics. Experimental results confirm this conjecture for a problem-specific problem representation (link-biased encoding), search operators (edge-set encoding), and initial solutions. We find that it is *not* the particular type of modern heuristic used that is relevant for high performance but instead the appropriate consideration of problem-specific knowledge for the design of the basic design elements.

References

Aarts EHL, Lenstra JK (eds) (1997) Local Search in Combinatorial Optimization. Wiley, New York

Aarts EHL, van Laarhoven PJM (1985) Statistical cooling: A general approach to combinatorial optimization problems. Philips Journal of Research 40:193–226

Aarts EHL, Korst JHM, van Laarhoven PJM (1997) Simulated annealing. In: Aarts EHL, Lenstra JK (eds) Local Search in Combinatorial Optimization, Discrete Mathematics and Optimization, Wiley-Interscience, Chichester, chap 4, pp 91–120

Abuali FN, Wainwright RL, Schoenefeld DA (1995) Determinant factorization: A new encoding scheme for spanning trees applied to the probabilistic minimum spanning tree problem. In: Eschelman (1995), pp 470–477

Ackley DH (1987) A connectionist machine for genetic hill climbing. Kluwer Academic, Boston

Ackoff R (1973) Science in the systems age: Beyond IE, OR and MS. Operations Research 21:661–671

Alander JT (2000) Indexed bibliography of genetic algorithms in economics. Tech. Rep. 94-1-ECO, University of Vaasa, Department of Information Technology and Production Economics

Alba E (2005) Parallel Metaheuristics: A New Class of Algorithms. Wiley, New York

Altenberg L (1994) The schema theorem and Price's theorem. In: Whitley and Vose (1994), pp 23–49

Altenberg L (1997) Fitness distance correlation analysis: An instructive counterexample. In: Bäck (1997), pp 57–64

Amberg A, Domschke W, Voß S (1996) Capacitated minimum spanning trees: Algorithms using intelligent search. Comb Opt: Theory and Practice 1:9–39

Angel E, Zissimopoulos V (1998a) Autocorrelation coefficient for the graph bipartitioning problem. Theoretical Computer Science 191:229–243

Angel E, Zissimopoulos V (1998b) On the quality of local search for the quadratic assignment problem. Discrete Appl Math 82(1–3):15–25

F. Rothlauf, *Design of Modern Heuristics*, Natural Computing Series, DOI 10.1007/978-3-540-72962-4, © Springer-Verlag Berlin Heidelberg 2011

Angel E, Zissimopoulos V (2000) On the classification of NP-complete problems in terms of their correlation coefficient. Discrete Appl Math 99(1–3):261–277

Angel E, Zissimopoulos V (2001) On the landscape ruggedness of the quadratic assignment problem. Theor Comput Sci 263(1–2):159–172

Angel E, Zissimopoulos V (2002) On the hardness of the quadratic assignment problem with metaheuristics. J Heuristics 8(4):399–414

Angeline PJ, Michalewicz Z, Schoenauer M, Yao X, Zalzala A, Porto W (eds) (1999) Proceedings of the 1999 IEEE Congress on Evolutionary Computation, IEEE Press, Piscataway

Arora S (1998) Polynomial-time approximation schemes for Euclidean TSP and other geometric problems. Journal of the ACM 45(5):753–782

Arora S, Barak B (2009) Computational complexity. Cambridge University Press

Arora S, Grigni M, Karger D, Klein P, Woloszyn A (1998) A polynomial-time approximation scheme for weighted planar graph TSP. In: Proceedings of the 9th Annual ACM-SIAM Symposium on Discrete Algorithms (SODA-98), ACM Press, New York, pp 33–41

Asoh H, Mühlenbein H (1994) On the mean convergence time of evolutionary algorithms without selection and mutation. In: Davidor et al (1994), pp 88–97

Bachmann PGH (1894) Die analytische Zahlentheorie. B. G. Teubner, Leipzig

Bäck T (1996) Evolutionary Algorithms in Theory and Practice. Oxford University Press, New York

Bäck T (ed) (1997) Proceedings of the Seventh International Conference on Genetic Algorithms, Morgan Kaufmann, San Francisco

Bäck T (1998) An overview of parameter control methods by self-adaptation in evolutionary algorithms. Fundamenta Informaticae 35:51–66

Bäck T, Schwefel HP (1995) Evolution strategies I: Variants and their computational implementation. In: Winter G, Périaux J, Galán M, Cuesta P (eds) Genetic Algorithms in Engineering and Computer Science, Wiley, Chichester, chap 6, pp 111–126

Bäck T, Fogel DB, Michalewicz Z (eds) (1997) Handbook of Evolutionary Computation. Institute of Physics Publishing and Oxford University Press, Bristol and New York

Bagchi S, Uckun S, Miyabe Y, Kawamura K (1991) Exploring problem-specific recombination operators for job shop scheduling. In: Belew and Booker (1991), pp 10–17

Bagley JD (1967) The behavior of adaptive systems which employ genetic and correlation algorithms. PhD thesis, University of Michigan

Balas E, Vazacopoulos A (1998) Guided local search with shifting bottleneck for job shop scheduling. Management Science 44:262–275

Baluja S (1994) Population-based incremental learning: A method for integrating genetic search based function optimization and competitive learning. Tech. Rep. CMU-CS-94-163, Carnegie Mellon University, Pittsburgh, PA

Baluja S, Davies S (1997) Using optimal dependency-trees for combinatorial optimization: Learning the structure of the search space. In: Proc. 14th International Conference on Machine Learning, Morgan Kaufmann, Burlington, pp 30–38

Bandyopadhyay S, Kargupta H, Wang G (1998) Revisiting the GEMGA: Scalable evolutionary optimization through linkage learning. In: Fogel (1998), pp 603–608

Banzhaf W, Langdon WB (2002) Some considerations on the reason for bloat. Genetic Programming and Evolvable Machines 3(1):81–91

Banzhaf W, Reeves CR (eds) (1998) Foundations of Genetic Algorithms 5, Morgan Kaufmann, San Francisco

Banzhaf W, Nordin P, Keller RE, Francone FD (1997) Genetic Programming. An Introduction. Morgan Kaufmann, Burlington

Banzhaf W, Daida J, Eiben AE, Garzon MH, Honavar V, Jakiela M, Smith RE (eds) (1999) Proceedings of the Genetic and Evolutionary Computation Conference, GECCO '99, Morgan Kaufmann, Burlington

Barnett L (1998) Ruggedness and neutrality: The NKp family of fitness landscapes. In: Adami C, Belew RK, Kitano H, Taylor CE (eds) Proceedings of the 6th International Conference on Artificial Life (ALIFE-98), MIT Press, Cambridge, pp 18–27

Bartz-Beielstein T (2006) Experimental Research in Evolutionary Computation. The New Experimentalism. Springer, Heidelberg

Battiti R, Protasi M (2001) Reactive local search for the maximum clique problem. Algorithmica 29(4):610–637

Battiti R, Tecchiolli G (1994) The reactive tabu search. ORSA J Comput 6(2):126–140

Baum EB (1986a) Iterated descent: A better algorithm for local search in combinatorial optimization problems. Tech. Rep., Caltech, Pasadena

Baum EB (1986b) Towards practical "neural" computation for combinatorial optimization problems. In: Denker JS (ed) Neural Networks for Computing, American Institute of Physics, New York, pp 53–58

Belew RK, Booker LB (eds) (1991) Proceedings of the Fourth International Conference on Genetic Algorithms, Morgan Kaufmann, Burlington

Belew RK, Vose MD (eds) (1996) Foundations of Genetic Algorithms 4, Morgan Kaufmann, San Francisco

Bellman R (1952) On the theory of dynamic programming. Proc Natl Acad Sci USA 38(8):716–719

Bellman R (1953) An introduction to the theory of dynamic programming. RAND Corporation, R-245

Bellman R (2003) Dynamic Programming. Dover, Mineda

Benson HY (2010a) Interior-point linear programming solvers. Tech. Rep., Drexel University, URL http://www.pages.drexel.edu/~hvb22/lpsolvers.pdf

Benson HY (2010b) Mixed integer nonlinear programming using interior-point methods. Optimization Methods and Software DOI 10.1080/105567810033799303

Bern MW (1988) Two probabilistic results on rectilinear Steiner trees. Algorithmica 3(2):191–204

Berry LTM, Murtagh BA, McMahon G (1995) Applications of a genetic-based algorithm for optimal design of tree-structured communication networks. In: Proceedings of the Regional Teletraffic Engineering Conference of the International Teletraffic Congress, Telkom South Africa, Pretoria, South Africa, pp 361–370

Berry LTM, Murtagh BA, McMahon G, Sugden S (1997) Optimization models for communication network design. In: Proceedings of the Fourth International Meeting, Decision Sciences Institute, Pitman, London, pp 67–70

Berry LTM, Murtagh BA, McMahon G, Sugden S, Welling L (1999) An integrated GA–LP approach to communication network design. Telecommunication Systems 12(2):265–280

Bertsekas DP, Tsitsiklis JN, Wu C (1997) Rollout algorithms for combinatorial optimization. J Heuristics 3(3):245–262

Bethke AD (1981) Genetic algorithms as function optimizers. PhD thesis, University of Michigan

Beyer HG, Deb K (2001) On self-adaptive features in real-parameter evolutionary algorithms. IEEE Trans Evolutionary Computation 5(3):250–270

Beyer HG, Schwefel HP (2002) Evolution strategies – A comprehensive introduction. Natural Computing 1(1):3–52

Beyer HG, Schwefel HP, Wegener I (2002) How to analyse evolutionary algorithms. Theor Comput Sci 287(1):101–130

Bierwirth C (1995) A generalized permutation approach to job shop scheduling with genetic algorithms. OR Spektrum 17:87–92

Bierwirth C, Mattfeld DC, Kopfer H (1996) On permutation representations for scheduling problems. In: Voigt et al (1996), pp 310–318

Bierwirth C, Mattfeld DC, Watson JP (2004) Landscape regularity and random walks for the job-shop scheduling problem. In: Gottlieb J, Raidl GR (eds) Evolutionary Computation in Combinatorial Optimization – EvoCOP 2004, Springer, Berlin, LNCS, vol 3004, pp 21–30

Biethahn J, Nissen V (eds) (1995) Evolutionary Algorithms in Management Applications. Springer, Berlin

Blum C, Roli A (2003) Metaheuristics in combinatorial optimization: Overview and conceptual comparison. ACM Comput Surv 35(3):268–308

Boese KD (1995) Cost versus distance in the traveling salesman problem. Tech. Rep. TR-950018, University of California at Los Angeles, Computer Science Department, Los Angeles

Bonissone P, Subbu R, Eklund N, Kiehl T (2006) Evolutionary algorithms + domain knowledge = real-world evolutionary computation. IEEE Transactions on Communications 10(3):256–280

Bonze IM, Grossmann W (1993) Optimierung – Theorie und Algorithmen. BI-Wissenschaftsverlag, Mannheim

Booker LB (1997) Binary strings. In: Bäck (1997), pp C3.3:1–C3.3:10

Boorstyn RR, Frank H (1977) Large-scale network topological optimization. IEEE Transactions on Communications 25:29–47

Borah M, Owens RM, Irwin MJ (1999) A fast and simple Steiner routing heuristic. Discrete Applied Mathematics 90:51–67

Bosman P (2003) Design and application of iterated density-estimation evolutionary algorithms. PhD thesis, Universiteit Utrecht, Utrecht, The Netherlands

Bosman PAN, Thierens D (2000) Expanding from discrete to continuous estimation of distribution algorithms: The IDEA. In: Schoenauer et al (2000), pp 767–776

Brameier M, Banzhaf W (2002) Explicit control of diversity and effective variation distance in linear genetic programming. In: Foster et al (2002), pp 162–171

Bredon G (1993) Topology and geometry, Graduate Texts in Mathematics, vol 139. Springer

Bronson R, Naadimuthu G (1997) Schaum's Outline of Operations Research, 2nd edn. McGraw-Hill

Burke EK, Kendall G (eds) (2005) Search Methodologies. Springer, New York

Buskes G, van Rooij A (1997) Topological Spaces: From Distance to Neighborhood. Undergraduate Texts in Mathematics, Springer, New York

Cahn RS (1998) Wide Area Network Design, Concepts and Tools for Optimization. Morgan Kaufmann, San Francisco

Caminiti S, Petreschi R (2005) String coding of trees with locality and heritability. In: Wang L (ed) Computing and Combinatorics, 11th Annual International Conference, COCOON 2005, Springer, Berlin, LNCS, vol 3595, pp 251–262

Caminiti S, Finocchi I, Petreschi R (2004) A unified approach to coding labeled trees. In: Farach-Colton M (ed) LATIN 2004: Theoretical Informatics, 6th Latin American Symposium, Springer, Berlin, LNCS, vol 2976, pp 339–348

Cantú-Paz E, Foster JA, Deb K, Davis D, Roy R, O'Reilly UM, Beyer HG, Standish R, Kendall G, Wilson S, Harman M, Wegener J, Dasgupta D, Potter MA, Schultz AC, Dowsland K, Jonoska N, Miller J (eds) (2003) Proceedings of the Genetic and Evolutionary Computation Conference, GECCO 2003, Springer, Heidelberg

Caruana RA, Schaffer JD (1988) Representation and hidden bias: Gray vs. binary coding for genetic algorithms. In: Laird L (ed) Proceedings of the Fifth International Workshop on Machine Learning, Morgan Kaufmann, San Francisco, pp 153–161

Caruana RA, Schaffer JD, Eshelman LJ (1989) Using multiple representations to improve inductive bias: Gray and binary coding for genetic algorithms. In: Segre AM (ed) Proceedings of the Sixth International Workshop on Machine Learning, Morgan Kaufmann, San Francisco, pp 375–378

Cayley A (1889) A theorem on trees. Quarterly Journal of Mathematics 23:376–378

Cerny V (1985) A thermodynamical approach to the travelling salesman problem: an efficient simulation algorithm. Journal of Optimization Theory and Applications 45:41–51

Chandra B, Karloff H, Tovey C (1994) New results on the old k-opt algorithm for the TSP. In: SODA '94: Proceedings of the Fifth Annual ACM-SIAM Symposium on Discrete Algorithms, Society for Industrial and Applied Mathematics, Philadelphia, pp 150–159

Chang SG, Gavish B (1993) Telecommunications network topological design and capacity expansion: formulations and algorithms. Telecommun Syst 1:99–131

Chao TH, Hsu YC (1994) Rectilinear Steiner tree construction by local and global refinement. IEEE Transactions on Computer-Aided Design of Integrated Circutits and Systems 13(3):303–309

Choi SS, Moon BR (2003) Normalization in genetic algorithms. In: Cantú-Paz et al (2003), pp 862–873

Choi SS, Moon BR (2008) Normalization for genetic algorithms with non-synonymously redundant encodings. IEEE Trans Evolutionary Computation 12(5):604–616

Choukhmane EA (1978) Une heuristique pour le problème de l'arbre de Steiner. RAIRO Rech Opér 12:207–212

Christensen S, Oppacher F (2001) What can we learn from no free lunch? In: Spector et al (2001), pp 1219–1226

Christofides N (1976) Worst-case analysis of a new heuristic for the traveling salesman problem. Tech. Rep. 388, Graduate School of Industrial Administration, Carnegie-Mellon University, Pittsburgh, PA

Church A (1936) An unsolvable problem of elementary number theory. American Journal of Mathematics 58:354–363

Chvátal V (1973) Edmonds polytopes and a hierarchy of combinatorial problems. Discrete Mathematics 4:305–337

Chvatal V (1983) Linear Programming. W. H. Freeman, New York

Coello Coello CA (1999) A survey of constraint handling techniques used with evolutionary algorithms. Technical report, Laboratorio Nacional de Informática Avanzada, Mexico

Coello Coello CA, Lamont GB, van Veldhuizen DA (2007) Evolutionary Algorithms for Solving Multi-Objective Problems, 2nd edn. Springer, New York

Cohoon JP, Hegde SU, Martin WN, Richards D (1988) Floorplan design using distributed genetic algorithms. In: IEEE International Conference on Computer Aided-Design, IEEE, Piscataway, pp 452–455

Collette Y, Siarry P (2004) Multiobjective Optimization: Principles and Case Studies. Springer, Heidelberg

Collins RJ, Jefferson DR (1991) Selection in massively parallel genetic algorithms. In: Belew and Booker (1991), pp 249–256

Contreras AA, Rowe JE, Stephens CR (2003) Coarse-graining in genetic algorithms: Some issues and examples. In: Cantú-Paz et al (2003), pp 874–885

Cook SA (1971) The complexity of theorem proving procedures. In: Proceedings of the Third ACM Symposium on Theory of Computing, ACM, New York, pp 151–158

Cormen TH, Leiserson CE, Rivest RL, Stein C (2001) Introduction to Algorithms. MIT Press, Cambridge

Cramer G (1750) Intr. à l'analyse des lignes courbes algèbriques (Introduction to the analysis of algebraic curves). Geneva

Cramer NL (1985) A representation for the adaptive generation of simple sequential programs. In: Grefenstette (1985), pp 183–187

Crescenzi P, Kann V (2003) A compendium of NP optimization problems. URL http://www.nada.kth.se/theory/compendium

Daida JM, Bertram R, Stanhope S, Khoo J, Chaudhary S, Chaudhri O, Polito J (2001) What makes a problem GP-hard? Analysis of a tunably difficult problem in genetic programming. Genetic Programming and Evolvable Machines 2(2):165–191

Dakin RJ (1965) A tree search algorithm for mixed integer programming problems. The Computer Journal 8(3):250–255

Dantzig GB (1949) Programming interdependent activities, II, mathematical model. Econometrica 17:200–211

Dantzig GB (1951) Maximization of a linear function of variables subject to linear inequalities. In: Koopmans T (ed) Activity Analysis of Production and Allocation, Wiley, New York, pp 339–347

Dantzig GB (1962) Linear Programming and Extensions. Princeton University Press, Princeton

Dantzig GB, Fulkerson DR, Johnson SM (1954) Solutions of a large-scale travelling salesman problem. Operations Research 2:393–410

Darwin C (1859) On the Origin of Species. John Murray, London

Davidor Y (1989) Epistasis variance – Suitability of a representation to genetic algorithms. Tech. Rep. No. CS89-25, Department of Applied Mathematics and Computer Science, The Weizmann Institute of Science, Rehovot, Israel

Davidor Y (1991) Epistasis variance: A viewpoint on GA-hardness. In: Rawlins (1991), pp 23–35

Davidor Y, Schwefel HP, Männer R (eds) (1994) Parallel Problem Solving from Nature – PPSN III, LNCS, vol 866, Springer, Berlin

Davis L (1985) Applying adaptive algorithms to epistatic domains. In: Joshi A (ed) Proc. of the 9th Int. Joint Conf. on Artificial Intelligence, Morgan Kaufmann, San Francisco, pp 162–164

Davis L, Orvosh D, Cox A, Qiu Y (1993) A genetic algorithm for survivable network design. In: Forrest (1993), pp 408–415

de Bonet JS, Isbell Jr CL, Viola P (1997) MIMIC: Finding optima by estimating probability densities. In: Mozer MC, Jordan MI, Petsche T (eds) Advances in Neural Information Processing Systems, MIT Press, Cambridge, vol 9, pp 424–430

De Jong KA (1975) An analysis of the behavior of a class of genetic adaptive systems. PhD thesis, University of Michigan, Ann Arbor, (University Microfilms No. 76-9381)

De Jong KA (2006) Evolutionary Computation. MIT Press, Cambridge

Deb K (2001) Multi-Objective Optimization using Evolutionary Algorithms. Wiley, Chichester

Deb K, Goldberg DE (1993a) Analyzing deception in trap functions. In: Whitley (1993), pp 93–108

Deb K, Goldberg DE (1993b) An investigation of niche and species formation in genetic function optimization. In: Forrest (1993), pp 42–50

Deb K, Goldberg DE (1994) Sufficient conditions for deceptive and easy binary functions. Annals of Mathematics and Artificial Intelligence 10:385–408

Deb K, Altenberg L, Manderick B, Bäck T, Michalewicz Z, Mitchell M, Forrest S (1997) Theoretical foundations and properties of evolutionary computations: fitness landscapes. In: Bäck et al (1997), pp B2.7:1–B2.7:25

Deb K, Anand A, Joshi D (2002) A computationally efficient evolutionary algorithm for real-parameter optimization. Evolutionary Computation 10(4):345–369

Deb K, Poli R, Banzhaf W, Beyer HG, Burke E, Darwen P, Dasgupta D, Floreano D, Foster J, Harman M, Holland O, Lanzi PL, Spector L, Tettamanzi A, Thierens D, Tyrrell A (eds) (2004) Proceedings of the Genetic and Evolutionary Computation Conference, GECCO 2004, Springer, Heidelberg

Denardo EV (2003) Dynamic programming: Models and applications. Dover, Mineda

Deo N, Micikevicius P (2001) Prüfer-like codes for labeled trees. Congressus Numerantium 151:65–73

Doerner K, Gendreau M, Greistorfer P, Gutjahr W, Hartl R, Reimann M (eds) (2007) Metaheuristics, Operations Research Computer Science Interfaces Series, vol 39. Springer, New York

Domschke W, Drexl A (2005) Einführung in Operations Research, 6th edn. Springer, Berlin

Doran J, Michie D (1966) Experiments with the graph traverser program. Proceedings of the Royal Society of London (A) 294:235–259

Dréo J, Pétrowski A, Siarry P, Taillard ED (2005) Metaheuristics for Hard Optimization. Methods and Case Studies. Springer, Berlin

Droste S, Wiesmann D (2002) On the design of problem-specific evolutionary algorithms. In: Ghosh A, Tsutsui S (eds) Advances in Evolutionary Computing: Theory and Applications, Springer, Berlin, pp 153–173

Droste S, Jansen T, Wegener I (2002) On the analysis of the $(1 + 1)$ evolutionary algorithm. Theoretical Computer Science 276(1–2):51–81

Dueck G, Scheuer T (1990) Threshold accepting: A general purpose optimization algorithm appearing superior to simulated annealing. J Comput Phys 90(1):161–175

Duin C, Voß S (1999) The pilot method: A strategy for heuristic repetition with application to the Steiner problem in graphs. Networks 34(3):181–191

Edelson W, Gargano ML (2000) Feasible encodings for GA solutions of constrained minimal spanning tree problems. In: Whitley et al (2000), p 754

Edelson W, Gargano ML (2001) Leaf constrained minimal spanning trees solved by a GA with feasible encodings. In: Wu AS (ed) Proceedings of the 2001 Genetic and Evolutionary Computation Conference Workshop Program, San Francisco, pp 268–271

Eiben AE, Smith JE (2010) Introduction to Evolutionary Computing, 2nd edn. Springer, Berlin

Elias D, Ferguson MJ (1974) Topological design of multipoint teleprocessing networks. IEEE Transactions on Communications 22:1753–1762

English TM (2000) Practical implications of new results in conservation of optimizer performance. In: Schoenauer et al (2000), pp 69–78

Ernst AT, Krishnamoorthy M, Storer RH (1999) Heuristic and exact algorithms for scheduling aircraft landings. Networks: An International Journal 34:229–241

Eschelman L (ed) (1995) Proceedings of the Sixth International Conference on Genetic Algorithms, Morgan Kaufmann, San Francisco

Eshelman LJ, Schaffer JD (1991) Preventing premature convergence in genetic algorithms by preventing incest. In: Belew and Booker (1991), pp 115–122

Etxeberria R, Larrañaga P (1999) Global optimization using Bayesian networks. In: Rodriguez AAO, Ortiz MRS, Hermida R (eds) Second Symposium on Artificial Intelligence. Distributions and Evolutionary Optimization. CIMAF 99, La Habana, pp 332–339

Evans JR (2006) Statistics, Data Analysis, and Decision Modeling, 3rd edn. Prentice Hall, Upper Saddle River

Falkenauer E (1998) Genetic Algorithms and Grouping Problems. Wiley, Chichester

Faroe O, Pisinger D, Zachariasen M (2003a) Guided local search for final placement in VLSI design. J Heuristics 9(3):269–295

Faroe O, Pisinger D, Zachariasen M (2003b) Guided local search for the three-dimensional bin-packing problem. INFORMS Journal on Computing 15(3):267–283

Feo TA, Resende MGC (1995) Greedy randomized adaptive search procedures. Journal of Global Optimization 6:109–133

Feo TA, Resende MGC, Smith SH (1994) A greedy randomized adaptive search procedure for maximum independent set. Operation Research 42(5):860–878

Fischer T (2007) Improved local search for large optimum communication spanning tree problems. In: MIC 2007 – 7th Metaheuristics International Conference

Fischer T, Merz P (2007) A memetic algorithm for the optimum communication spanning tree problem. In: Bartz-Beielstein T, Aguilera MJB, Blum C, Naujoks B, Roli A, Rudolph G, Sampels M (eds) Hybrid Metaheuristics, 4th International Workshop, HM 2007, Dortmund, Germany, October 8–9, 2007, Proceedings, Springer, Berlin, LNCS, vol 4771, pp 170–184, DOI http://dx.doi.org/10.1007/978-3-540-75514-2_13

Fogel DB (1995) Evolutionary Computation. IEEE Press, Piscataway

Fogel DB (1997) Real-valued vectors. In: Bäck et al (1997), pp C3.2:2–C3.2:5

Fogel DB (ed) (1998) Proceedings of 1998 IEEE International Conference on Evolutionary Computation, IEEE Service Center, Piscataway

Fogel DB, Attikiouzel Y (eds) (1995) Proceedings of the 1995 IEEE International Conference on Evolutionary Computation, IEEE Service Center, Piscataway

Fogel DB, Stayton LC (1994) On the effectiveness of crossover in simulated evolutionary optimization. BioSystems 32:171–182

Fogel L (1999) Intelligence through Simulated Evolution: Forty Years of Evolutionary Programming. Wiley, New York

Fogel L, Owens A, Walsh M (1966) Artificial Intelligence through Simulated Evolution. Wiley, New York

Fonlupt C, Hao JK, Lutton E, Ronald EMA, Schoenauer M (eds) (1999) Proceedings of Artificial Evolution: Fifth European Conference, LNCS, vol 1829, Springer, Berlin

Forrest S (ed) (1993) Proceedings of the Fifth International Conference on Genetic Algorithms, Morgan Kaufmann, San Francisco

Forsgren A, Gill PE, Wright MH (2002) Interior methods for nonlinear optimization. SIAM Review 44(4):525–597

Fortnow L (2009) The status of the P versus NP problem. Commun ACM 52(9):78–86

Foster JA, Lutton E, Miller J, Ryan C, Tettamanzi AGB (eds) (2002) Proceedings of the Fifth European Conference on Genetic Programming (EuroGP-2002), LNCS, vol 2278, Springer, Berlin

Foulds LR (1983) The heuristic problem-solving approach. Journal of the Operational Research Society 10(34):927–934

Gale JS (1990) Theoretical Population Genetics. Unwin Hyman, London

Gallagher M, Frean M (2005) Population-based continuous optimization, probabilistic modelling and mean shift. Evolutionary Computation 13(1):29–42

Garey MR, Johnson DS (1979) Computers and Intractability: A Guide to the Theory of NP-completeness. W. H. Freeman, New York

Gargano ML, Edelson W, Koval O (1998) A genetic algorithm with feasible search space for minimal spanning trees with time-dependent edge costs. In: Koza JR, Banzhaf W, Chellapilla K, Deb K, Dorigo M, Fogel DB, Garzon MH, Goldberg DE, Iba H, Riolo RL (eds) Genetic Programming 98, Morgan Kaufmann, San Francisco, p 495

Gaube T, Rothlauf F (2001) The link and node biased encoding revisited: Bias and adjustment of parameters. In: Boers EJW, Cagnoni S, Gottlieb J, Hart E, Lanzi PL, Raidl GR, Smith RE, Tijink H (eds) Applications of Evolutionary Computing: Proc. EvoWorkshops 2001. Springer, Berlin, pp 1–10

Gavish B, Li CL, Simchi-Levi D (1992) Analysis of heuristics for the design of tree networks. Annals of Operations Research 36:77–86

Gelernter H (1963) Realization of a geometry-theorem proving machine. In: Feigenbaum EA, Feldman J (eds) Computers and thought, McGraw-Hill, New York, pp 134–152, published 1959 in Proceedings of International Conference on Information Processing, Unesco House

Gelly S, Teytaud O, Bredeche N, Schoenauer M (2005) A statistical learning theory approach of bloat. In: Beyer HG, O'Reilly UM, Arnold DV, Banzhaf W, Blum C, Bonabeau EW, Cantu-Paz E, Dasgupta D, Deb K, Foster JA, de Jong ED, Lipson H, Llora X, Mancoridis S, Pelikan M, Raidl GR, Soule T, Tyrrell AM, Watson JP, Zitzler E (eds) Proceedings of the Genetic and Evolutionary Computation Conference, GECCO 2005, ACM Press, New York, pp 1783–1784

Gen M, Li Y (1999) Spanning tree-based genetic algorithms for the bicriteria fixed charge transportation problem. In: Angeline et al (1999), pp 2265–2271

Gen M, Zhou G, Takayama M (1998) A comparative study of tree encodings on spanning tree problems. In: Fogel (1998), pp 33–38

Gendreau M (2003) An introduction to tabu search. In: Glover F, Kochenberger GA (eds) Handbook of Metaheuristics, Kluwer, Alphen aan den Rijn, pp 37–54

Gendreau M, Potvin JY (2005) Metaheuristics in combinatorial optimization. Annals OR 140(1):189–213. DOI http://dx.doi.org/10.1007/s10479-005-3971-7

Gerrits M, Hogeweg P (1991) Redundant coding of an NP-complete problem allows effective genetic algorithm search. In: Schwefel HP, Männer R (eds) Parallel Problem Solving from Nature – PPSN I, Springer, Berlin, LNCS, vol 496, pp 70–74

Gill PE, Murray W, Saunders MA, Tomlin JA, Wright MH (1986) On projected Newton barrier methods for linear programming and an equivalence to Karmarkar's projective method. Mathematical Programming 36:183–209

Glover F (1977) Heuristics for integer programming using surrogate constraints. Decision Sciences 8(1):156–166

Glover F (1986) Future paths for integer programming and links to artificial intelligence. Computers & OR 13(5):533–549, DOI http://dx.doi.org/10.1016/0305-0548(86)90048-1

Glover F (1990) Tabu search – part II. ORSA Journal on Computing 2(1):4–32

Glover F (1994) Genetic algorithms and scatter search – unsuspected potentials. Statistics And Computing 4(2):131–140

Glover F (1997) A template for scatter search and path relinking. In: Hao JK, Lutton E, Ronald EMA, Schoenauer M, Snyers D (eds) Proceedings of Artificial Evolution: Fourth European Conference, Berlin, LNCS, vol 1363, pp 1–51

Glover F, Kochenberger GA (eds) (2003) Handbook of Metaheuristics. Kluwer, Boston

Glover F, Laguna M (1997) Tabu Search. Kluwer, Boston

Goldberg DE (1987) Simple genetic algorithms and the minimal, deceptive problem. In: Davis L (ed) Genetic Algorithms and Simulated Annealing, Morgan Kaufmann, San Mateo, chap 6, pp 74–88

Goldberg DE (1989a) Genetic algorithms and Walsh functions: Part I, a gentle introduction. Complex Systems 3(2):129–152

Goldberg DE (1989b) Genetic algorithms and Walsh functions: Part II, deception and its analysis. Complex Systems 3(2):153–171

Goldberg DE (1989c) Genetic algorithms in search, optimization, and machine learning. Addison-Wesley, Reading

Goldberg DE (1990) Real-coded genetic algorithms, virtual alphabets, and blocking. IlliGAL Report No. 90001, University of Illinois at Urbana-Champaign, Urbana, IL

Goldberg DE (1991a) Genetic algorithm theory. Fourth International Conference on Genetic Algorithms Tutorial, unpublished manuscript

Goldberg DE (1991b) Real-coded genetic algorithms, virtual alphabets, and blocking. Complex Systems 5(2):139–167

Goldberg DE (1992) Construction of high-order deceptive functions using low-order Walsh coefficients. Annals of Mathematics and Artificial Intelligence 5:35–48

Goldberg DE (2002) The design of Innovation. Series on Genetic Algorithms and Evolutionary Computation, Kluwer, Dordrecht

Goldberg DE, Deb K (1991) A comparative analysis of selection schemes used in genetic algorithms. Foundations of Genetic Algorithms 1:69–93

Goldberg DE, Lingle, Jr R (1985) Alleles, loci, and the traveling salesman problem. In: Grefenstette (1985), pp 154–159

Goldberg DE, Segrest P (1987) Finite Markov chain analysis of genetic algorithms. In: Grefenstette (1987), pp 1–8

Goldberg DE, Korb B, Deb K (1989) Messy genetic algorithms: Motivation, analysis, and first results. Complex Systems 3(5):493–530

Goldberg DE, Deb K, Clark JH (1992) Genetic algorithms, noise, and the sizing of populations. Complex Systems 6:333–362

Goldberg DE, Deb K, Kargupta H, Harik G (1993a) Rapid, accurate optimization of difficult problems using fast messy genetic algorithms. In: Forrest (1993), pp 56–64

Goldberg DE, Deb K, Thierens D (1993b) Toward a better understanding of mixing in genetic algorithms. Journal of the Society of Instrument and Control Engineers 32(1):10–16

Gomory RE (1958) Outline of an algorithm for integer solutions to linear programs. Bulletin of the American Mathematical Society 64:275–278

Gomory RE (1960) Solving linear programming problems in integers. In: Bellman R, Hall, Jr M (eds) Combinatorial Analysis, Symposia in Applied Mathematics X, American Mathematical Society, Providence, pp 211–215

Gomory RE (1963) An algorithm for integer solutions to linear programs. In: Graves RL, Wolfe P (eds) Recent Advances in Mathematical Programming. McGraw-Hill, New York, pp 269–302

Gomory RE, Hu TC (1961) Multi-terminal network flows. In: SIAM Journal on Applied Math, vol 9, pp 551–570

Gottlieb J (1999) Evolutionary algorithms for constrained optimization problems. PhD thesis, Technische Universität Clausthal, Institut für Informatik, Clausthal, Germany

Gottlieb J, Raidl GR (1999) Characterizing locality in decoder-based EAs for the multidimensional knapsack problem. In: Fonlupt et al (1999), pp 38–52

Gottlieb J, Raidl GR (2000) The effects of locality on the dynamics of decoder-based evolutionary search. In: Whitley et al (2000), pp 283–290

Gottlieb J, Julstrom BA, Raidl GR, Rothlauf F (2001) Prüfer numbers: A poor representation of spanning trees for evolutionary search. In: Spector et al (2001), pp 343–350

Grahl J, Minner S, Rothlauf F (2005) Behaviour of $UMDA_c$ with truncation selection on monotonous functions. In: Corne D, Michalewicz Z, McKay B, Eiben G, Fogel D, Fonseca C, Greenwood G, Raidl G, Tan KC, Zalzala A (eds) Proceedings of 2005 IEEE Congress on Evolutionary Computation, IEEE Press, Piscataway, vol 3, pp 2553–2559

Grahl J, Radtke A, Minner S (2007) Fitness landscape analysis of dynamic multi-item lot-sizing problems with limited storage. In: Günther HO, Mattfeld DC, Suhl L (eds) Management logistischer Netzwerke – Entscheidungsunterstützung, Informationssysteme und OR-Tools, Physica, Heidelberg, pp 257–277

Gray F (1953) Pulse code communications. U.S. Patent 2632058

Grefenstette JJ (ed) (1985) Proceedings of an International Conference on Genetic Algorithms and Their Applications, Lawrence Erlbaum, Hillsdale

Grefenstette JJ (ed) (1987) Proceedings of the Second International Conference on Genetic Algorithms, Lawrence Erlbaum, Hillsdale

Grefenstette JJ, Gopal R, Rosmaita BJ, Van Gucht D (1985) Genetic algorithms for the traveling salesman problem. In: Grefenstette (1985), pp 160–168

Grötschel M, Jünger M, Reinelt G (1984) A cutting plane algorithm for the linear ordering problem. Operations Research 32:1195–1220

Grünert T, Irnich S (2005) Optimierung im Transport 1: Grundlagen. Shaker, Aachen

Gu J (1992) Efficient local search for very large-scale satisfiability problems. SIGART Bulletin 3(1):8–12

Gu J, Purdom P, Franco J, Wah B (1996) Algorithms for satisfiability (sat) problem: A survey. DIMACS Volume Series on Discrete Mathematics and Theoretical Computer Science 35:19–151

Hamming R (1980) Coding and Information Theory. Prentice Hall, Englewood Cliffs

Handa H (2006) Fitness function for finding out robust solutions on time-varying functions. In: Keijzer M, Cattolico M, Arnold D, Babovic V, Blum C, Bosman P, Butz MV, Coello Coello C, Dasgupta D, Ficici SG, Foster J, Hernandez-Aguirre A, Hornby G, Lipson H, McMinn P, Moore J, Raidl G, Rothlauf F, Ryan C, Thierens D (eds) Proceedings of the Genetic and Evolutionary Computation Conference, GECCO 2006, ACM Press, New York, pp 1195–1200

Hansen N, Ostermeier A (2001) Completely derandomized self-adaptation in evolution strategies. Evolutionary Computation 9(2):159–195

Hansen P, Mladenović N (2001) Variable neighborhood search – principles and applications. European Journal of Operational Research 130:449–467

Hansen P, Mladenović N (2003) Variable neighborhood search. In: Glover F, Kochenberger G (eds) Handbook of Metaheuristics, Kluwer Academic, Dordrecht, pp 145–184

Harel D, Rosner R (1992) Algorithmics: The spirit of computing, 2nd edn. Addison-Wesley, Reading

Harik G (1999) Linkage learning via probabilistic modeling in the ECGA. IlliGAL Report No. 99010, University of Illinois at Urbana-Champaign, Urbana, IL

Harik GR (1997) Learning gene linkage to efficiently solve problems of bounded difficulty using genetic algorithms. PhD thesis, University of Michigan, Ann Arbor

Harik GR, Goldberg DE (1996) Learning linkage. In: Belew and Vose (1996), pp 247–262

Harik GR, Lobo FG, Goldberg DE (1998) The compact genetic algorithm. In: Fogel (1998), pp 523–528

Hartl DL, Clark AG (1997) Principles of population genetics, 3rd edn. Sinauer, Sunderland

Hartmanis J, Stearns RE (1965) On the computational complexity of algorithms. Transactions of the American Mathematical Society 117:285–306

Heckendorn RB, Whitley D, Rana S (1996) Nonlinearity, Walsh coefficients, hyperplane ranking and the simple genetic algorithm. In: Belew and Vose (1996), pp 181–201

Heckendorn RB, Rana S, Whitley D (1999) Polynomial time summary statistics for a generalization of MAXSAT. In: Banzhaf et al (1999), pp 281–288

Held M, Karp RM (1962) A dynamic programming approach to sequencing problems. Journal of the Society for Industrial and Applied Mathematics 10(1):196–210

Helsgaun K (2000) An effective implementation of the Lin-Kernighan traveling salesman heuristic. European Journal of Operational Research 126(1):106–130

Henderson D, Jacobson SH, Johnson AW (2003) The theory and practice of simulated annealing. In: Glover F, Kochenberger GA (eds) Handbook of Metaheuristics, Kluwer, Dordrecht, pp 287–319

Heroelen WS (1972) Heuristic programming in operations management. Die Unternehmung 26:213–231

Hillier FS, Lieberman GJ (2002) Introduction to Operations Research, 7th edn. McGraw-Hill, New York

Ho JM, Vijayan G, Wong CK (1990) New algorithms for the rectilinear Steiner tree problem. IEEE Transactions on Computer-Aided Design 9:185–193

Hoai NX, McKay RI, Essam DL (2006) Representation and structural difficulty in genetic programming. IEEE Trans Evolutionary Computation 10(2):157–166

Hocaoglu C, Sanderson AC (1995) Evolutionary speciation using minimal representation size clustering. In: McDonnell JR, Reynolds RG, Fogel DB (eds) Proc. 4th Annual Conf. Evolutionary Programming, MIT Press, Cambridge, pp 187–203

Hochbaum DS (ed) (1996) Approximation Algorithms for NP-hard Problems. PWS, Boston

Hochbaum DS, Shmoys DB (1987) Using dual approximation algorithms for scheduling problems: Theoretical and practical results. Journal of the ACM 34(1):144–162

Höhn C, Reeves CR (1996a) Are long path problems hard for genetic algorithms? In: Voigt et al (1996), pp 134–143

Höhn C, Reeves CR (1996b) The crossover landscape for the onemax problem. In: Alander JT (ed) Proc. of the Second Nordic Workshop on Genetic Algorithms and their Applications (2NWGA), Department of Information Technology and Production Economics, University of Vaasa, Vaasa, Finnland, pp 27–43

Holland JH (1975) Adaptation in natural and artificial systems. University of Michigan Press, Ann Arbor

Hoos HH, Stützle T (2004) Stochastic Local Search: Foundations & Applications. Morgan Kaufmann, San Francisco

Horn J (1995) Genetic algorithms, problem difficulty, and the modality of fitness landscapes. IlliGAL Report No. 95004, University of Illinois at Urbana-Champaign, Urbana, IL

Horn J, Goldberg DE (1998) Toward a control map for niching. In: Banzhaf and Reeves (1998), pp 287–310

Hu TC (1974) Optimum communication spanning trees. SIAM Journal on Computing 3(3):188–195

Huynen M, Stadler P, Fontana W (1996) Smoothness within ruggedness: The role of neutrality in adaptation. In: Proceedings of the National Academy of Sciences of the USA, 1993, National Academy of Sciences, Washington, D.C., pp 397–401

Hwang FK (1976) On Steiner minimal trees with rectilinear distance. SIAM J Applied Math 1:104–114

Ibaraki T, Nonobe K, Yagiura M (eds) (2005) Metaheuristics: Progress as Real Problem Solvers. Springer, New York

Ibarra OH, Kim CE (1975) Fast approximation algorithms for the knapsack and sum of subset problems. Journal of the ACM 22(4):463–468

Igel C (1998) Causality of hierarchical variable length representations. In: Fogel (1998), pp 324–329

Igel C, Chellapilla K (1999) Investigating the influence of depth and degree of genotypic change on fitness in genetic programming. In: Banzhaf et al (1999), pp 1061–1068

Inayoshi H, Manderick B (1994) The weighted graph bi-partitioning problem: A look at GA performance. In: Davidor et al (1994), pp 617–625

Jansen T, Wegener I (2002) The analysis of evolutionary algorithms – A proof that crossover really can help. Algorithmica 34(1):47–66

Jansen T, Wegener I (2005) Real royal road functions–where crossover provably is essential. Discrete Applied Mathematics 149(1–3):111–125, DOI http://dx.doi.org/10.1016/j.dam.2004.02.019

Johnson DS (1974) Approximation algorithms for combinatorial problems. Journal of Computer and System Sciences 9(3):256–278

Johnson DS (1990) Local optimization and the traveling salesman problem. In: Paterson M (ed) Automata, Languages and Programming, 17th International Colloquium, ICALP90, Warwick University, England, July 16–20, 1990, Proceedings, Springer, Berlin, LNCS, vol 443, pp 446–461

Johnson DS, Lenstra JK, Kan AHGR (1978) The complexity of the network design problem. Networks 8:279–285

Jones DR, Beltramo MA (1991) Solving partitioning problems with genetic algorithms. In: Belew and Booker (1991), pp 442–449

Jones T (1995a) Evolutionary algorithms, fitness landscapes and search. PhD thesis, University of New Mexico, Albuquerque, NM

Jones T (1995b) One operator, one landscape. Tech. Rep. 95-02-025, Santa Fe Institute

Jones T, Forrest S (1995) Fitness distance correlation as a measure of problem difficulty for genetic algorithms. In: Eschelman (1995), pp 184–192

Julstrom BA (1999) Redundant genetic encodings may not be harmful. In: Banzhaf et al (1999), p 791

Julstrom BA (2001) The blob code: A better string coding of spanning trees for evolutionary search. In: Wu AS (ed) Proceedings of the 2001 Genetic and Evolutionary Computation Conference Workshop Program, San Francisco, pp 256–261

Julstrom BA (2002) A scalable genetic algorithm for the rectilinear Steiner problem. In: Fogel DB, El-Sharkawi MA, Yao X, Greenwood G, Iba H, Marrow P, Shackleton M (eds) Proceedings of 2002 IEEE Congress on Evolutionary Computation, IEEE Press, Piscataway, pp 1169–1173

Kahng AB, Robins G (1992) On performance bounds for a class of rectilinear Steiner tree heuristics in arbitrary dimension. IEEE Transactions on Computer-Aided Design 11(11):1462–1465

Kangshen S, Crossley JN, Lun AW (eds) (2000) The Nine Chapters on the Mathematical Art: Companion and Commentary. Oxford University Press, Oxford

Kantorovich LV (1939) The Mathematical Method of Production Planning and Organization. Leningrad University Press, Leningrad, in Russian

Kargupta H, Deb K, Goldberg DE (1992) Ordering genetic algorithms and deception. In: Männer and Manderick (1992), pp 47–56

Karmarkar N (1984) A new polynomial-time algorithm for linear programming. Combinatorica 4:273–395

Karp RM (1972) Reducibility among combinatorial problems. In: Miller RE, Thatcher JW (eds) Complexity of Computer Computations, Plenum, New York, pp 85–103

Kauffman SA (1989) Adaptation on rugged fitness landscapes. In: Stein DL (ed) Lectures in the sciences of complexity SFI Studies in the Sciences of Complexity, Addison-Wesley, New York, pp 527–618

Kauffman SA (1993) Origins of order: Self-organization and selection in evolution. Oxford University Press, Oxford

Kauffman SA, Levin S (1987) Towards a general theory of adaptive walks on rugged landscapes. J Theor Biol 128:11–45

Keller RE, Banzhaf W (1996) Genetic programming using genotype-phenotype mapping from linear genomes into linear phenotypes. In: Koza JR, Goldberg DE, Fogel DB, Riolo RL (eds) Proceedings of First Annual Conference on Genetic Programming, MIT Press, Cambridge, pp 116–122

Kelner JA, Spielman DA (2006) A randomized polynomial-time simplex algorithm for linear programming. In: STOC '06: Proceedings of the thirty-eighth annual ACM symposium on Theory of computing, ACM Press, New York, pp 51–60

Kershenbaum A (1993) Telecommunications network design algorithms. McGraw-Hill, New York

Kershenbaum A, Chou W (1974) A unified algorithm for designing multidrop teleprocessing networks. IEEE Transactions on Communications 22:1762–1772

Kershenbaum A, Boorstyn RR, Oppenheim R (1980) Second-order greedy algorithms for centralized teleprocessing network design. IEEE Transactions on Communications 28:1835–1838

Khachian LG (1979) A polynomial algorithm in linear programming. Dokl Akad Nauk SSSR 244(5):1093–1096, english translation in Soviet Math. Dokl. 20:191–194, 1979

Kim JR, Gen M (1999) Genetic algorithm for solving bicriteria network topology design problem. In: Angeline et al (1999), pp 2272–2279

Kimura M (1962) On the probability of fixation of mutant genes in a population. Genetics 47:713–719

Kimura M (1964) Diffusion models in population genetics. J Appl Prob 1:177–232

Kirkpatrick S, Toulouse G (1985) Configuration space analysis for traveling salesman problems. Journal de Physique 46:1277–1292

Kirkpatrick S, Gelatt CD, Vecchi MP (1983) Optimization by simulated annealing. Science 220:671–680

Klee V, Minty GJ (1972) How good is the simplex algorithm? In: Shisha, O (ed) Inequalities – III, Academic Press, New York and London, pp 159–175

Kleinau P, Thonemann UW (2004) Deriving inventory-control policies with genetic programming. OR Spectrum 26(4):521–546, DOI http://dx.doi.org/10.1007/s00291-004-0159-5

Knuth DE (1998) The Art of Computer Programming. Volume 3: Sorting and Searching, 2nd edn. Addison-Wesley, New York

Kolarov K (1997) Landscape ruggedness in evolutionary algorithms. In: Bäck T, Michalewicz Z, Yao X (eds) Proceedings of the Fourth International Conference on Evolutionary Computation, IEEE Service Center, Piscataway, pp 19–24

Koza JR (1992) Genetic Programming: On the programming of computers by natural selection. MIT Press, Cambridge

Koza JR (1994) Genetic Programming II: Automatic Discovery of Reusable Programs. MIT Press, Cambridge

Koza JR, Keane MA, Bennett FHI, Andre D (1999) Genetic Programming III: Darwinian invention and problem solving. Morgan Kaufmann, San Francisco

Koza JR, Keane MA, Streeter MJ, Mydlowec W, Yu J, Lanza G (2005) Genetic Programming IV: Routine human-competitive machine intelligence. Springer, New York

Krishnamoorthy M, Ernst AT (2001) Comparison of algorithms for the degree constrained minimum spanning tree. Journal of Heuristics 7:587–611

Kruskal JB (1956) On the shortest spanning subtree of a graph and the travelling salesman problem. Proc Amer Math Soc 7:48–50

Kuehn AA, Hamburger MJ (1963) A heuristic program for locating warehouses. Management Science 9(9):643–666

Kursawe F (1999) Grundlegende empirische Untersuchungen der Parameter von Evolutionsstrategien – Metastrategien. PhD thesis, University of Dortmund

Laguna M, Martí R (2003) Scatter search: Methodology and Implementations in C. Operations Research Computer Science Interfaces Series, vol 24. Kluwer, Dordrecht

Land A, Doig A (1960) An automatic method of solving discrete programming problems. Econometrika 28(3):497–520

Landau E (1974) Handbuch der Lehre von der Verteilung der Primzahlen, 3rd edn. Chelsea Publishing Company, Providence, reprinted from the First Edition, 1909

Langdon W, Poli R (1997) Fitness causes bloat: Mutation. In: Koza J (ed) Late Breaking Papers at GP'07, Stanford Bookstore, pp 132–140

Langdon WB, Poli R (2002) Foundations of Genetic Programming. Springer, Berlin

Larrañaga P, Lozano JA (eds) (2001) Estimation of Distribution Algorithms: A New Tool for Evolutionary Computation. Springer, Berlin

Larrañaga P, Etxeberria R, Lozano JA, Pea JM (1999) Optimization by learning and simulation of Bayesian and Gaussian networks. Tech. Rep. EHU-KZAA-IK-4/99, Intelligent Systems Group, Dept. of Computer Science and Artificial Intelligence, University of the Basque Country

Lawler EL (1979) Fast approximation algorithms for knapsack problems. Math Oper Res 4(4):339–356

Lee EK (2002) Branch-and-bound methods. In: Pardalos PM, Resende MGC (eds) Handbook of Applied Optimization, Oxford University Press, Oxford, pp 53–65

Lee EK, Mitchell JE (2001) Branch-and-bound methods for integer programming. In: Floudas CA, Pardalos PM (eds) Encyclopedia of Optimization, Kluwer Academic, Dordrecht

Leung Y, Gao Y, Xu ZB (1997) Degree of population diversity – a perspective on premature convergence in genetic algorithms and its Markov chain analysis. IEEE Transactions on Neural Networks 8(5):1165–1176

Lewontin RC (1974) The Genetic Basis of Evolutionary Change. No. 25 in Columbia Biological Series, Columbia University Press, New York

Li Y (2001) An effective implementation of a direct spanning tree representation in GAs. In: Boers EJW, Cagnoni S, Gottlieb J, Hart E, Lanzi PL, Raidl GR, Smith RE, Tijink H (eds) Applications of Evolutionary Computing: Proc. EvoWorkshops 2001, Springer, Berlin, pp 11–19

Li Y, Gen M, Ida K (1998) Fixed charge transportation problem by spanning tree-based genetic algorithm. Beijing Mathematics 4(2):239–249

Liepins GE, Vose MD (1990) Representational issues in genetic optimization. Journal of Experimental and Theoretical Artificial Intelligence 2:101–115

Liepins GE, Vose MD (1991) Polynomials, basis sets, and deceptiveness in genetic algorithms. Complex Systems 5(1):45–61

Lin S (1965) Computer solutions of the travelling salesman problem. Bell Systems Technical Journal 44:2245–2269

Lin S, Kernighan BW (1973) An effective heuristic algorithm for the traveling-salesman problem. Operations Research 21(2):498–516

Lobo FG, Goldberg DE, Pelikan M (2000) Time complexity of genetic algorithms on exponentially scaled problems. In: Whitley et al (2000), pp 151–158

Lohmann R (1993) Structure evolution and incomplete induction. Biol Cybern 69(4):319–326

Lourenco HR, Martin OC, Stützle T (2001) Iterated local search. In: Glover F, Kochenberger G (eds) Handbook of Metaheuristics, Kluwer, Dordrecht, pp 321–353

Luke S, Panait L (2006) A comparison of bloat control methods for genetic programming. Evolutionary Computation 14(3):309–344, DOI http://dx.doi.org/10.1162/evco.2006.14.3.309

Maclaurin C (1748) Treatise of algebra. London

Mahfoud SW (1995) A comparison of parallel and sequential niching methods. In: Eschelman (1995)

Manderick B, de Weger M, Spiessens P (1991) The genetic algorithm and the structure of the fitness landscape. In: Belew and Booker (1991), pp 143–150

Männer R, Manderick B (eds) (1992) Parallel Problem Solving from Nature – PPSN II, Springer, Berlin

Maros I, Mitra G (1996) Simplex algorithms. In: Beasley JE (ed) Advances in Linear and Integer Programming, Oxford Lecture Series in Mathematics and its Applications, vol 4, Clarendon, Oxford, pp 1–46

Martin O, Otto SW (1996) Combining simulated annealing with local search heuristics. Annals of Operations Research 63:57–75

Martin O, Otto SW, Felten EW (1991) Large-step Markov chains for the traveling salesman problem. Complex Systems 5:299–326

Mason A (1995) A non-linearity measure of a problem's crossover suitability. In: Fogel and Attikiouzel (1995), pp 68–73

Mattfeld DC (1996) Evolutionary Search and the Job Shop: Investigations on Genetic Algorithms for Production Scheduling. Physica, Heidelberg

Mendel G (1866) Versuche über Pflanzen-Hybriden. In: Verhandlungen des naturforschenden Vereins, Naturforschender Verein zu Brünn, Brünn, vol 4, pp 3–47

Mendes AS, França PM, Moscato P (2002) Fitness landscapes for the total tardiness single machine scheduling problem. Neural Network World: An International Journal on Neural and Mass-Parallel Computing and Information Systems 12(2):165–180

Merz P, Freisleben B (2000a) Fitness landscape analysis and memetic algorithms for the quadratic assignment problem. IEEE Transactions on Evolutionary Computation 4(4):337–352

Merz P, Freisleben B (2000b) Fitness landscapes, memetic algorithms, and greedy operators for graph bipartitioning. Evolutionary Computation 8(1):61–91

Metropolis N, Rosenbluth A, Rosenbluth MN, Teller A, Teller E (1958) Equations of state calculations by fast computing machines. J Chem Phys 21:1087–1092

Michalewicz Z (1996) Genetic algorithms + data structures = evolution programs. Springer, New York

Michalewicz Z, Fogel DB (2004) How to Solve It: Modern Heuristics, 2nd edn. Springer, Berlin

Michalewicz Z, Janikow CZ (1989) Handling constraints in genetic algorithm. In: Schaffer (1989), pp 151–157

Michalewicz Z, Schoenauer M (1996) Evolutionary computation for constrained parameter optimization problems. Evolutionary Computation 4(1):1–32

Michalewicz Z, Deb K, Schmidt M, Stidsen TJ (1999) Towards understanding constraint-handling methods in evolutionary algorithms. In: Angeline et al (1999), pp 581–588

Miller JF, Thomson P (2000) Cartesian genetic programming. In: Poli R, Banzhaf W, Langdon WB, Miller J, Nordin P, Fogarty TC (eds) Genetic Programming: Third European Conference, Springer, Berlin, LNCS, vol 1802, pp 121–132

Mills P, Tsang EPK (2000) Guided local search for solving SAT and weighted MAX-SAT problems. Journal of Automated Reasoning 24:205–223

Mitchell M (1996) An introduction to genetic algorithms. MIT Press, Cambridge

Mitchell M, Forrest S, Holland JH (1992) The royal road for genetic algorithms: Fitness landscapes and GA performance. In: Varela FJ, Bourgine P (eds) Proc. of the First European Conference on Artificial Life, MIT Press, Cambridge, pp 245–254

Mitchell T (1982) Generalization as search. Artificial Intelligence 18(2):203–226

Mittelmann H (2010) Decision tree for optimization software. Tech. Rep., Arizona State University, http://plato.asu.edu/guide.html

Mladenovic N, Hansen P (1997) Variable neighborhood search. Computers & OR 24(11):1097–1100, DOI http://dx.doi.org/10.1016/S0305-0548(97)00031-2

Moraglio A (2007) Towards a geometric unification of evolutionary algorithms. PhD thesis, Department of Computer Science, University of Essex

Moraglio A, Poli R (2004) Topological interpretation of crossover. In: Deb et al (2004), pp 1377–1388

Moraglio A, Kim YH, Yoon Y, Moon BR (2007) Geometric crossovers for multiway graph partitioning. Evolutionary Computation 15(4):445–474, DOI http://dx.doi.org/10.1162/evco.2007.15.4.445

Mühlenbein H, Mahnig T (1999) FDA – a scalable evolutionary algorithm for the optimization of additively decomposed functions. Evolutionary Computation 7(4):353–376

Mühlenbein H, Paaß G (1996) From recombination of genes to the estimation of distributions i. binary parameters. In: Voigt et al (1996), pp 178–187

Nagylaki T (1992) Introduction to Theoretical Population Genetics, Biomathematics, vol 21. Springer, Berlin

Naphade KS, Wu SD, Storer RH (1997) Problem space search algorithms for resource-constrained project scheduling. Annals of Operations Research 70:307–326

Naudts B, Suys D, Verschoren A (1997) Epistasis as a basic concept in formal landscape analysis. In: Bäck (1997), pp 65–72

Neumann F (2007) Expected runtimes of a simple evolutionary algorithm for the multi-objective minimum spanning tree problem. European Journal of Operational Research 181(3):1620–1629

Neumann F, Wegener I (2007) Randomized local search, evolutionary algorithms, and the minimum spanning tree problem. Theor Comput Sci 378(1):32–40, DOI http://dx.doi.org/10.1016/j.tcs.2006.11.002

Neville EH (1953) The codifying of tree-structure. Proceedings of the Cambridge Philosophical Society 49:381–385

Newell A (1969) Heuristic programming: Ill-structured problems. In: Aronofsky J (ed) Progress in Operations Research, vol 3, Wiley, New York, pp 60–413

Nissen V (1997) Einführung in evolutionäre Algorithmen: Optimierung nach dem Vorbild der Evolution. Vieweg, Wiesbaden

Nordin P, Banzhaf W (1995) Complexity compression and evolution. In: Eschelman (1995), pp 310–317

Oei CK (1992) Walsh function analysis of genetic algorithms of non-binary strings. Master's thesis, University of Illinois at Urbana-Champaign, Department of Computer Science, Urbana

Oliver IM, Smith DJ, Holland JRC (1987) A study of permutation crossover operators on the traveling salesman problem. In: Grefenstette (1987)

O'Neill M (2006) URL http://www.grammatical-evolution.org

O'Neill M, Ryan C (2001) Grammatical evolution. IEEE Transactions on Evolutionary Computation 5(4):349–358

O'Neill M, Ryan C (2003) Grammatical Evolution: Evolutionary Automatic Programming in an Arbitrary Language, Genetic Programming, vol 4. Kluwer Academic

O'Reilly UM (1997) Using a distance metric on genetic programs to understand genetic operators. In: Late breaking papers at the 1997 Genetic Programming Conference, Stanford University, pp 199–206

Orponen P, Mannila H (1987) On approximation preserving reductions: complete problems and robust measures. Tech. Rep. C-1987-28, Dept. of Computer Science, University of Helsinki, Finland

Osman IH, Kelly JP (eds) (1996) Metaheuristics theory and applications. Kluwer Academic, Boston

Osman IH, Laporte G (1996) Metaheuristics: A bibliography. Annals of Operations Research 63:513–623

Padberg M, Rinaldi G (1991) A branch-and-cut algorithm for the resolution of large-scale symmetric traveling salesman problems. SIAM Review 33(1):60–100

Palmer CC (1994) An approach to a problem in network design using genetic algorithms. PhD thesis, Polytechnic University, Brooklyn, NY

Palmer CC, Kershenbaum A (1994) Representing trees in genetic algorithms. In: Proceedings of the First IEEE Conference on Evolutionary Computation, IEEE Service Center, Piscataway, vol 1, pp 379–384

Papadimitriou C (1994) Computational complexity. Addison-Wesley, Reading

Papadimitriou CH, Steiglitz K (1978) Some examples of difficult traveling salesman problems. Oper Res 26(3):434–443

Papadimitriou CH, Steiglitz K (1982) Combinatorial Optimization: Algorithms and Complexity. Prentice Hall, Englewood Cliffs

Papadimitriou CH, Vempala S (2000) On the approximability of the traveling salesman problem. In: Proceedings of the 32nd Annual ACM Symposium on Theory of Computing, STOC 2000 (Portland, Oregon, May 21–23, 2000), ACM, New York, pp 126–133

Papadimitriou CH, Yannakakis M (1991) Optimization, approximation, and complexity classes. J Comput System Sci 43:425–440

Papadimitriou CH, Yannakakis M (1993) The traveling salesman problem with distances one and two. Mathematics of Operations Research 18:1–11

Paulden T, Smith DK (2006) From the dandelion code to the rainbow code: A class of bijective spanning tree representations with linear complexity and bounded locality. IEEE Transactions on Evolutionary Computation 10(2):124–144

Pearl J (1984) Heuristics: Intelligent Search Strategies for Computer Problem Solving. Addison-Wesley, Reading

Peleg D, Reshef E (1998) Deterministic polylog approximation for minimum communication spanning trees. In: Larsen KG, Skyum S, Winskel G (eds) Automata, Languages and Programming, Proc. of the 25th Intl. Colloquium (ICALP'98), Springer, Berlin, LNCS, vol 1443, pp 670–682

Pelikan M (2006) Hierarchical Bayesian Optimization Algorithm: Toward a New Generation of Evolutionary Algorithms. Studies in Fuzziness and Soft Computing, Springer, New York

Pelikan M, Mühlenbein H (1999) The bivariate marginal distribution algorithm. In: Roy R, Furuhashi T, Chawdhry PK (eds) Advances in Soft Computing – Engineering Design and Manufacturing, Springer, London, pp 521–535

Pelikan M, Goldberg DE, Cantú-Paz E (1999a) BOA: The Bayesian optimization algorithm. IlliGAL Report No. 99003, University of Illinois at Urbana-Champaign, Urbana, IL

Pelikan M, Goldberg DE, Lobo F (1999b) A survey of optimization by building and using probabilistic models. IlliGAL Report No. 99018, University of Illinois at Urbana-Champaign, Urbana, IL

Picciotto S (1999) How to encode a tree. PhD thesis, University of California, San Diego, USA

Polya G (1945) How to solve it. Penguin Books Ltd., London

Polya G (1948) How to solve it. Princeton Univesity Press, Princeton, NJ

Porto VW, Saravanan N, Waagen D, Eiben AE (eds) (1998) Evolutionary Programming VII. LNCS, vol 1447. Springer, Berlin

Power DJ (2002) Decision Support Systems: Concepts and Resources for Managers. Quorum, Westport

Prim R (1957) Shortest connection networks and some generalizations. Bell System Technical Journal 36:1389–1401

Prüfer H (1918) Neuer Beweis eines Satzes über Permutationen. Archiv für Mathematik und Physik 27:742–744

Puchta M, Gottlieb J (2002) Solving car sequencing problems by local optimization. In: Cagnoni S, Gottlieb J, Hart E, Middendorf M, Raidl GR (eds) Applications of Evolutionary Computing, Springer, Berlin, LNCS, vol 2279, pp 131–140

Radcliffe NJ (1991a) Equivalence class analysis of genetic algorithms. Complex Systems 5(2):183–205

Radcliffe NJ (1991b) Forma analysis and random respectful recombination. In: Rawlins (1991), pp 222–229

Radcliffe NJ (1992) Non-linear genetic representations. In: Männer and Manderick (1992), pp 259–268

Radcliffe NJ (1993) Genetic set recombination. In: Whitley (1993), pp 203–219

Radcliffe NJ (1994) The algebra of genetic algorithms. Annals of Maths and Artificial Intelligence 10:339–384

Radcliffe NJ (1997) Theoretical foundations and properties of evolutionary computations: Schema processing. In: Bäck et al (1997), pp B2.5:1–B2.5:10

Radcliffe NJ, Surry PD (1994) Fitness variance of formae and performance prediction. In: Whitley and Vose (1994), pp 51–72

Radcliffe NJ, Surry PD (1995) Fundamental limitations on search algorithms: Evolutionary computing in perspective. In: van Leeuwen J (ed) Computer Science Today: Recent Trends and Developments, LNCS, vol 1000, Springer, Berlin, pp 275–291, DOI http://dx.doi.org/10.1007/BFb0015249

Raidl GR (2001) Various instances of optimal communication spanning tree problems. Personal communciation

Raidl GR, Gottlieb J (2005) Empirical analysis of locality, heritability and heuristic bias in evolutionary algorithms: A case study for the multidimensional knapsack problem. Evolutionary Computation 13(4):441–475

Raidl GR, Julstrom BA (2000) A weighted coding in a genetic algorithm for the degree-constrained minimum spanning tree problem. In: Carroll J, Damiani E, Haddad H, Oppenheim D (eds) Proceedings of the 2000 ACM Symposium on Applied Computing, ACM Press, New York, pp 440–445

Raidl GR, Julstrom BA (2003) Edge-sets: An effective evolutionary coding of spanning trees. IEEE Transactions on Evolutionary Computation 7(3):225–239

Rana SB, Whitley LD (1998) Genetic algorithm behavior in the MAXSAT domain. In: Eiben AE, Bäck T, Schoenauer M, Schwefel HP (eds) Parallel Problem Solving from Nature – PPSN V, Springer, Berlin, LNCS, vol 1498, pp 785–794

Rana SB, Heckendorn RB, Whitley LD (1998) A tractable Walsh analysis of SAT and its implications for genetic algorithms. In: Proceedings of the 15th National Conference on Artificial Intelligence (AAAI-98) and of the 10th Conference on Innovative Applications of Artificial Intelligence (IAAI-98), AAAI Press, Menlo Park, pp 392–397

Rawlins GJE (ed) (1991) Foundations of Genetic Algorithms. Morgan Kaufmann, San Mateo

Rayward-Smith VJ (ed) (1998) Applications of Modern Heuristic Methods. Nelson Thornes, Cheltenham

Rayward-Smith VJ, Osman IH, Reeves CR, Smith GD (eds) (1996) Modern Heuristic Search Methods. Wiley, New York

Rechenberg I (1965) Cybernetic solution path of an experimental problem. Tech. Rep. 1122, Royal Aircraft Establishment, Library Translation, Farnborough, Hants., UK

Rechenberg I (1973a) Bionik, Evolution und Optimierung. Naturwissenschaftliche Rundschau 26(11):465–472

Rechenberg I (1973b) Evolutionsstrategie: Optimierung technischer Systeme nach Prinzipien der biologischen Evolution. Friedrich Frommann Verlag, Stuttgart-Bad Cannstatt

Rechenberg I (1994) Evolutionsstrategie'94, Werkstatt Bionik und Evolutionstechnik, vol 1. Friedrich Frommann Verlag (Günther Holzboog KG), Stuttgart

Reed J, Toombs R, Barricelli NA (1967) Simulation of biological evolution and machine learning: I. Selection of self-reproducing numeric patterns by data processing machines, effects of hereditary control, mutation type and crossing. Journal of Theoretical Biology 17:319–342

Reeves CR (ed) (1993) Modern heuristic techniques for combinatorial problems. Blackwell Scientific, Oxford

Reeves CR (1999a) Fitness landscapes and evolutionary algorithms. In: Fonlupt et al (1999), pp 3–20

Reeves CR (1999b) Landscapes, operators and heuristic search. Annals of Operational Research 86:473–490

Reeves CR, Rowe JE (2003) Genetic Algorithms: Principles and Perspectives. Kluwer, Dordrecht

Reeves CR, Wright C (1994) An experimental design perspective on genetic algorithms. In: Whitley and Vose (1994), pp 7–22

Reidys CM, Stadler PF (2002) Combinatorial landscapes. SIAM Review 44(1):3–54, DOI http://dx.doi.org/10.1137/S0036144501395952

Resende MG, de Sousa JP (eds) (2003) Metaheuristics: Computer Decision-Making. Springer, Berlin

Reshef E (1999) Approximating minimum communication cost spanning trees and related problems. Master's thesis, Feinberg Graduate School of the Weizmann Institute of Science, Rehovot 76100, Israel

Ribeiro C, Hansen P (eds) (2001) Essays and surveys in metaheuristics. Kluwer Academic, Dordrecht

Richards D (1989) Fast heuristic algorithms for rectilinear Steiner trees. Algorithmica 4:191–207

Robins G, Zelikovsky A (2005) Tighter bounds for graph Steiner tree approximation. SIAM J Discret Math 19(1):122–134, DOI http://dx.doi.org/10.1137/S0895480101393155

Romanycia MHJ, Pelletier FJ (1985) What is a heuristic? Computational Intelligence 1(2):47–58

Ronald S, Asenstorfer J, Vincent M (1995) Representational redundancy in evolutionary algorithms. In: Fogel and Attikiouzel (1995), pp 631–636

Rosenkrantz DJ, Stearns RE, Lewis PMI (1977) An analysis of several heuristics for the traveling salesman problem. SIAM J Comput 6(3):563–581

Rothlauf F (2002) Representations for Genetic and Evolutionary Algorithms, 1st edn. No. 104 in Studies on Fuzziness and Soft Computing, Springer, Heidelberg

Rothlauf F (2006) Representations for Genetic and Evolutionary Algorithms, 2nd edn. Springer, Heidelberg

Rothlauf F (2009a) On optimal solutions for the optimal communication spanning tree problem. Operations Research 57(2):413–425

Rothlauf F (2009b) On the bias and performance of the edge-set encoding. IEEE Transactions on Evolutionary Computation 13(3):486–499

Rothlauf F (2009c) A problem-specific and effective encoding for metaheuristics for the minimum communication spanning tree problem. INFORMS Journal on Computing 21(4):575–584

Rothlauf F, Goldberg DE (1999) Tree network design with genetic algorithms - an investigation in the locality of the prüfernumber encoding. In: Brave S, Wu AS (eds) Late Breaking Papers at the Genetic and Evolutionary Computation Conference 1999, Omni Press, Orlando, pp 238–244

Rothlauf F, Goldberg DE (2000) Prüfernumbers and genetic algorithms: A lesson on how the low locality of an encoding can harm the performance of GAs. In: Schoenauer et al (2000), pp 395–404

Rothlauf F, Goldberg DE (2003) Redundant representations in evolutionary computation. Evolutionary Computation 11(4):381–415

Rothlauf F, Goldberg DE, Heinzl A (2002) Network random keys – A tree network representation scheme for genetic and evolutionary algorithms. Evolutionary Computation 10(1):75–97

Rowe JE, Whitley LD, Barbulescu L, Watson JP (2004) Properties of Gray and binary representations. Evolutionary Computation 12(1):46–76

Rudolph G (1996) Convergence properties of evolutionary algorithms. PhD thesis, Universität Dortmund, Dortmund

Russell S, Norvig P (2002) Artificial Intelligence a Modern Approach, 2nd edn. AI, Prentice Hall

Ryan C (1999) Automatic Re-engineering of Software Using Genetic Programming. Kluwer Academic, Amsterdam

Ryan C, O'Neill M (1998) Grammatical evolution: A steady state approach. In: Late Breaking Papers, Genetic Programming 1998, pp 180–185

Salustowicz R, Schmidhuber J (1997) Probabilistic incremental program evolution. Evolutionary Computation 5(2):123–141

Sastry K, Goldberg DE (2001) Modeling tournament selection with replacement using apparent added noise. IlliGAL Report No. 2001014, University of Illinois at Urbana-Champaign, Urbana, IL

Sastry K, Goldberg DE (2002) Analysis of mixing in genetic algorithms: A survey. Tech. Rep. 2002012, IlliGAL, Department of General Engineering, University of Illinois at Urbana-Champaign

Schaffer JD (ed) (1989) Proceedings of the Third International Conference on Genetic Algorithms, Morgan Kaufmann, Burlington

Schneeweiß C (2003) Distributed Decision Making. Springer, Berlin

Schnier T (1998) Evolved representations and their use in computational creativity. PhD thesis, University of Sydney, Department of Architectural and Design Science

Schnier T, Yao X (2000) Using multiple representations in evolutionary algorithms. In: Fonseca C, Kim JH, Smith A (eds) Proceedings of 2000 IEEE Congress on Evolutionary Computation, IEEE Press, Piscataway, pp 479–486

Schoenauer M, Deb K, Rudolph G, Yao X, Lutton E, Merelo JJ, Schwefel HP (eds) (2000) Parallel Problem Solving from Nature – PPSN VI, Springer, Berlin

Schraudolph NN, Belew RK (1992) Dynamic parameter encoding for genetic algorithms. Machine Learning 9:9–21

Schumacher C (2000) Fundamental limitations of search. PhD thesis, University of Tennessee, Department of Computer Science, Knoxville, TN

Schumacher C, Vose MD, Whitley LD (2001) The no free lunch and problem description length. In: Spector et al (2001), pp 565–570

Schwefel HP (1965) Kybernetische Evolution als Strategie der experimentellen Forschung in der Strömungstechnik. Master's thesis, Technische Universität Berlin

Schwefel HP (1968) Experimentelle Optimierung einer Zweiphasendüse. Bericht 35, AEG Forschungsinstitut Berlin, Projekt MHD-Staustahlrohr

Schwefel HP (1975) Evolutionsstrategie und numerische Optimierung. PhD thesis, Technical University of Berlin

Schwefel HP (1977) Numerische Optimierung von Computer-Modellen mittels der Evolutionsstrategie. Birkhäuser, Basel

Schwefel HP (1981) Numerical Optimization of Computer Models. Wiley, Chichester

Schwefel HP (1995) Evolution and Optimum Seeking. Wiley, New York

Sebald AV, Chellapilla K (1998a) On making problems evolutionarily friendly – part 1: Evolving the most convenient representations. In: Porto et al (1998), pp 271–280

Sebald AV, Chellapilla K (1998b) On making problems evolutionarily friendly – part 2: Evolving the most convenient coordinate systems within which to pose (and solve) the given problem. In: Porto et al (1998), pp 281–290

Selkow SM (1977) The tree-to-tree editing problem. Information Processing Letters 6(6):184–186

Selman B, Levesque H, Mitchell D (1992) A new method for solving hard satisfiability problems. In: Swartout W (ed) Proceedings of the 10th National Conference on Artificial Intelligence, MIT Press, Cambridge, pp 440–446

Sendhoff B, Kreutz M, von Seelen W (1997a) Causality and the analysis of local search in evolutionary algorithms. Tech. Rep., Institut für Neuroinformatik, Ruhr-Universität Bochum

Sendhoff B, Kreutz M, von Seelen W (1997b) A condition for the genotype-phenotype mapping: Causality. In: Bäck (1997), pp 73–80

Shaefer CG (1987) The ARGOT strategy: adaptive representation genetic optimizer technique. In: Grefenstette (1987), pp 50–58

Shaefer CG, Smith JS (1990) The ARGOT strategy II: Combinatorial optimizations. Tech. Rep. RL90-1, Thinking Machines Inc.

Shang YW, Qiu YH (2006) A note on the extended Rosenbrock function. Evolutionary Computation 14(1):119–126

Shipman R (1999) Genetic redundancy: Desirable or problematic for evolutionary adaptation? In: Dobnikar A, Steele NC, Pearson DW, Albrecht RF (eds) Proceedings of the 4th International Conference on Artificial Neural Networks and Genetic Algorithms (ICANNGA), Springer, Berlin, pp 1–11

Shor NZ (1970) Utilization of the operation of space dilatation in the minimization of convex functions. Cybernetics and System Analysis 6:7–15

Siarry P, Michalewicz Z (eds) (2008) Advances in Metaheuristics for Hard Optimization. Natural Computing Series, Springer, Berlin

Silver EA (2004) An overview of heuristic solution methods. Journal of the Operational Research Society 55(9):936–956

Simon HA, Newell A (1958) Heuristic problem solving: The next advance in operations research. Operations Research 6:1–10

Simonoff JS (1996) Smoothing Methods in Statistics. Springer, Berlin

Sinclair MC (1995) Minimum cost topology optimisation of the COST 239 European optical network. In: Pearson DW, Steele NC, Albrecht RF (eds) Proceedings of the 1995 International Conference on Artificial Neural Nets and Genetic Algorithms, Springer, New York, pp 26–29

Smale S (1983) On the average number of steps in the simplex method of linear programming. Mathematical Programming 27:241–262

Smith AE, Coit DW (1997) Penalty functions. In: Bäck et al (1997), pp C5.2:1–6

Smith SF (1980) A learning system based on genetic adaptive algorithms. PhD thesis, University of Pittsburgh

Soto M, Ochoa A, Acid S, de Campos LM (1999) Introducing the polytree approximation of distribution algorithm. In: Rodriguez AAO, Ortiz MRS, Hermida R (eds) Second Symposium on Artificial Intelligence. Adaptive Systems. CIMAF 99, La Habana, pp 360–367

Soule T (2002) Exons and code growth in genetic programming. In: Foster et al (2002), pp 142–151

Spector L, Goodman E, Wu A, Langdon WB, Voigt HM, Gen M, Sen S, Dorigo M, Pezeshk S, Garzon M, Burke E (eds) (2001) Proceedings of the Genetic and Evolutionary Computation Conference, GECCO 2001, Morgan Kaufmann, San Francisco

Stadler PF (1992) Correlation in landscapes of combinatorial optimization problems. Europhys Lett 20:479–482

Stadler PF (1995) Towards a theory of landscapes. In: Lopéz-Peña R, Capovilla R, García-Pelayo R, Waelbroeck H, Zertuche F (eds) Complex Systems and Binary Networks, Springer, Berlin, Lecture Notes in Physics, vol 461, pp 77–163

Stadler PF (1996) Landscapes and their correlation functions. J Math Chem 20:1–45

Stadler PF, Schnabl W (1992) The landscape of the traveling salesman problem. Physics Letters A 161:337–344

Steitz W, Rothlauf F (2010) Solving OCST problems with problem-specific guided local search. In: Pelikan M, Branke J (eds) Proceedings of the Genetic and Evolutionary Computation Conference, GECCO 2010, ACM Press, New York, pp 301–302

Stephens CR, Waelbroeck H (1999) Schemata evolution and building blocks. Evolutionary Computation 7:109–124

Storer RH, Wu SD, Vaccari R (1992) New search spaces for sequencing problems with application to job shop scheduling. Management Science 38(10):1495–1509

Storer RH, Wu SD, Vaccari R (1995) Problem and heuristic space search strategies for job shop scheduling. ORSA Journal on Computing 7(4):453–467

Streeter MJ (2003) Two broad classes of functions for which a no free lunch result does not hold. In: Cantú-Paz (2003), pp 1418–1430

Stützle T (1999) Iterated local search for the quadratic assignment problem. Tech. Rep. AIDA-99-03, FG Intellektik, FB Informatik, TU Darmstadt

Surry PD (1998) A prescriptive formalism for constructing domain-specific evolutionary algorithms. PhD thesis, University of Edinburgh, Edinburgh

Surry PD, Radcliffe N (1996) Formal algorithms + formal representations = search strategies. In: Voigt et al (1996), pp 366–375

Suzuki J (1995) A Markov chain analysis on simple genetic algorithms. IEEE Transactions on Systems, Man, and Cybernetics 25(4):655–659

Syswerda G (1989) Uniform crossover in genetic algorithms. In: Schaffer (1989), pp 2–9

Taha HA (2002) Operations Research: An Introduction, 7th edn. Prentice Hall, Englewood Cliffs

Tai KC (1979) The tree-to-tree correction problem. Journal of the ACM 26(3):422–433

Taillard ÉD (1991) Robust taboo search for the quadratic assignment problem. Parallel Computing 17(4–5):443–455

Taillard ÉD (1995) Comparison of iterative searches for the quadratic assignment problem. Location Science 3:87–105

Taillard ÉD, Gambardella LM, Gendreau M, Potvin JY (2001) Adaptive memory programming: A unified view of metaheuristics. European Journal of Operational Research 135(1):1–16

Thierens D (1995) Analysis and design of genetic algorithms. PhD thesis, Katholieke Universiteit Leuven, Leuven, Belgium

Thierens D (1999) Scalability problems of simple genetic algorithms. Evolutionary Computation 7(4):331–352

Thierens D, Goldberg DE, Pereira ÂG (1998) Domino convergence, drift, and the temporal-salience structure of problems. In: Fogel (1998), pp 535–540

Tonge FM (1961) The use of heuristic programming in management science. Management Science 7:231–237

Turban E, Aronson JE, Liang TP (2004) Decision Support Systems and Intelligent Systems, 7th edn. Prentice Hall, Englewood Cliffs

Turing A (1936) On computable numbers with an application to the Entscheidungsproblem. Proc. London Mathematical Society 2(42):230–265

Tzschoppe C, Rothlauf F, Pesch HJ (2004) The edge-set encoding revisited: On the bias of a direct representation for trees. In: Deb et al (2004), pp 1174–1185

van Laarhoven PJM, Aarts EHL (1988) Simulated Annealing: Theory and Applications. Kluwer, Dordrecht

Vazirani VV (2003) Approximation Algorithms. Springer, Berlin

Voigt HM, Ebeling W, Rechenberg I, Schwefel HP (eds) (1996) Parallel Problem Solving from Nature – PPSN IV, Springer, Berlin

Vose MD (1993) Modeling simple genetic algorithms. In: Whitley (1993), pp 63–73

Vose MD (1999) The simple genetic algorithm: foundations and theory. MIT Press, Cambridge

Vose MD, Wright AH (1998a) The simple genetic algorithm and the Walsh transform: Part I, theory. Evolutionary Computation 6(3):253–273

Vose MD, Wright AH (1998b) The simple genetic algorithm and the Walsh transform: Part II, the inverse. Evolutionary Computation 6(3):275–289

Voß S (2001) Capacitated minimum spanning trees. In: Floudas CA, Pardalos PM (eds) Encyclopedia of Optimization, vol 1, Kluwer, Boston, pp 225–235

Voß S, Martello S, Osman IH, Roucairol C (eds) (1999) Metaheuristics: Advances and Trends in Local Search Paradigms for Optimization. Kluwer, Boston

Voudouris C (1997) Guided local search for combinatorial optimization problems. Phd thesis, Dept. of Computer Science, University of Essex, Colchester, UK

Voudouris C, Tsang E (1995) Guided local search. Tech. Rep. CSM-247, University of Essex, Colchester, UK

Voudouris C, Tsang E (1999) Guided local search. Europ J Oper Res 113(2):469–499

Watanabe S (1969) Knowing and guessing – A formal and quantitative study. Wiley, New York

Weicker K, Weicker N (1998) Locality vs. randomness – dependence of operator quality on the search state. In: Banzhaf and Reeves (1998), pp 289–308

Weinberger E (1990) Correlated and uncorrelated fitness landscapes and how to tell the difference. Biological Cybernetics 63:325–336

Weyl H (1935) Elementare Theorie der konvexen Polyeder. Comment Math Helv 7:290–306

Whigham PA, Dick G (2010) Implicitly controlling bloat in genetic programming. IEEE Transactions on Evolutionary Computation 14(2):173–190, DOI http://dx.doi.org/10.1109/TEVC.2009.2027314

Whitley LD (1989) The GENITOR algorithm and selection pressure: Why rank-based allocation. In: Schaffer (1989), pp 116–121

Whitley LD (ed) (1993) Foundations of Genetic Algorithms 2, Morgan Kaufmann, San Mateo

Whitley LD (1997) Permutations. In: Bäck et al (1997), pp C3.3:114–C3.3:20

Whitley LD (2002) Evaluating evolutionary algorithms. Tutorial Program at Parallel Problem Solving from Nature (PPSN 2002)

Whitley LD, Rowe J (2005) Gray, binary and real valued encodings: Quad search and locality proofs. In: Wright AH, Vose MD, De Jong KA, Schmitt LM (eds) Foundations of Genetic Algorithms 8, LNCS, vol 3469, Springer, Berlin, pp 21–36

Whitley LD, Rowe J (2008) Focused no free lunch theorems. In: Keijzer M, Antoniol G, Congdon CB, Deb K, Doerr B, Hansen N, Holmes JH, Hornby GS, Howard D, Kennedy J, Kumar S, Lobo FG, Miller JF, Moore J, Neumann F, Pelikan M, Pollack J, Sastry K, Stanley K, Stoica A, Talbi EG, Wegener I (eds) Proceedings of the Genetic and Evolutionary Computation Conference, GECCO 2008, ACM Press, New York, pp 811–818

Whitley LD, Rowe JE (2006) Subthreshold-seeking local search. Theor Comput Sci 361(1):2–17, DOI http://dx.doi.org/10.1016/j.tcs.2006.04.008

Whitley LD, Vose MD (eds) (1994) Foundations of Genetic Algorithms 3, Morgan Kaufmann, San Mateo

Whitley LD, Watson J (2005) Complexity theory and the no free lunch theorem. In: Burke EK, Kendall G (eds) Search Methodologies, Springer, pp 317–339, DOI http://dx.doi.org/10.1007/0-387-28356-0_11

Whitley LD, Goldberg DE, Cantú-Paz E, Spector L, Parmee L, Beyer HG (eds) (2000) Proceedings of the Genetic and Evolutionary Computation Conference, GECCO 2000, Morgan Kaufmann, San Francisco

Whitley LD, Bush K, Rowe JE (2004) Subthreshold-seeking behavior and robust local search. In: Deb et al (2004), pp 282–293

Wiest JD (1966) Heuristic programs for decision making. Harvard Business Review 44(5):129–143

Williams HP (1999) Model Building in Mathematical Programming, 4th edn. Wiley, Chichester

Winston WL (1991) Operations research: Applications and algorithms, 4th edn. Duxbury Press, Pacific Grove

Wolpert DH, Macready WG (1995) No free lunch theorems for search. Tech. Rep. No. SFI-TR-95-02-010, Santa Fe Institute, Santa Fe, NM

Wolpert DH, Macready WG (1997) No free lunch theorems for optimization. IEEE Trans on Evolutionary Computation 1(1):67–82

Wright MH (2005) The interior-point revolution in optimization: History, recent developments, and lasting consequences. BAMS: Bulletin of the American Mathematical Society 42:39–56

Wright S (1932) The roles of mutation, inbreeding, crossbreeding, and selection in evolution. In: Jones DF (ed) Proceedings of the Sixth International Congress of Genetics, Brooklyn Botanic Garden, Brooklyn, vol 1, pp 356–366

Wright SJ (1997) Primal-Dual Interior-Point Methods. Society for Industrial and Applied Mathematics

Wu BY, Chao KM (2004) Spanning Trees and Optimization Problems. Discrete Mathematics and Its Applications, Chapman & Hall/CRC, Boca Raton

Yannakakis M (1994) On the approximation of maximum satisfiability. J Algorithms 17(3):475–502

Yudin D, Nemirovskii A (1976) Information complexity and efficient methods for the solution of convex extremal problems. Matekon 13(2):3–25

Zhang K, Shasha D (1989) Simple fast algorithms for the editing distance between trees and related problems. SIAM Journal on Computing 18(6):1245–1262

Zhang W (1999) State-Space Search: Algorithms, Complexity, Extensions, and Applications. Springer, Berlin

Zhang W (2004) Configuration landscape analysis and backbone guided local search. part I: Satisfiability and maximum satisfiability. Artificial Intelligence 158(1):1–26

Zhou G, Gen M (1997) Approach to the degree-constrained minimum spanning tree problem using genetic algorithms. Engineering Design & Automation 3(2):156–165

Nomenclature

$*$	don't care symbol, page 41
α_i	ith coefficient of the polynomial decomposition of x, page 39
$\mathbf{e}_n$	vector that contains 1 in the nth column and 0 elsewhere, page 39
χ	cardinality, page 112
$\delta(H)$	defining length of schema H, page 41
γ	step size, page 49
λ	regularization parameter (GLS), page 137
$\mathscr{A}_f(x, S_f)$	algorithm that calculates the objective value of a solution x using the set of parameters S_X, page 25
$\mathscr{A}_X(x, S_X)$	algorithm that checks whether solution x is feasible using the set of parameters S_X, page 25
$\mathscr{N}_i(0, 1)$	normally distributed one-dimensional random variable with expectation zero and standard deviation one, page 142
$\nabla f(x)$	gradient of $f(x)$, page 46
$\Phi(d_m^y)$	performance function, page 98
Φ_g	genotype space, page 110
Φ_p	phenotype space, page 110
ψ_j	jth Walsh function, page 40
$\rho()$	approximation ratio, page 88
$\rho()$	autocorrelation function, page 35
ρ_{FDC}	fitness distance correlation coefficient, page 32
σ	permutation, page 99
$\sigma()$	standard deviation, page 32
$\wedge$	bitwise logical AND, page 40
$\zeta()$	ratio of successful to all search steps (ES), page 143
A	search algorithm, page 98
B_ε	open ball with radius ε, page 18
b_{ij}	biases, page 207
$bc(x)$	number of ones in x, page 40
c	crossover point, page 118
c_{fd}	covariance of f and d_{opt}, page 32

D set of decision variables, page 36, 37

d maximal depth of a search tree, page 61

$d(x,y)$ distance between x and y, page 16

d^g distance between genotypes, page 161

d^p distance between phenotypes, page 161

d_m locality of a representation, page 161

$d_m^x(i)$ ith solution generated by an algorithm, page 98

$d_{i,j}$ distance between i and j, page 32

d_{max} maximum distance, page 21

d_{min} minimum distance, page 21

d_{opt} distance between solution and optimal solution, page 32

$diam\,G$ diameter of graph G, page 21

E set of vertex pairs or edges, page 61

e_{uv}^i indicates whether an edge from u to v exists in T_i, page 187

F function set (genetic programming), page 153

$f()$ evaluation or objective function, page 14

f_g representation, page 110

f_p mapping of phenotypes to fitness, page 110

$f_s(H)$ fitness of schema H, page 42

G graph, page 61

$g()$ inequality constraints, page 15

G_L graph describing a fitness landscape, page 21

H schema, page 41

$h()$ equality constraints, page 15

$H(f(x))$ Hessian matrix of $f(x)$, page 47

$h(x)$ evaluation function, page 64

I set of problem instances, page 14

$I_i()$ indicator function (GLS), page 137

$iter$ maximum number of search steps, page 210

$iter_{term}$ maximum number of no-improvement search steps, page 210

k depth of parse trees, page 153

k_{max} order of a deceptive problem, page 43

l problem size, length of solution, page 36

l_{corr} correlation length, page 36

M mating pool, page 148

m number of edges, page 187

N population size, page 147

n input size, problem size, number of nodes, page 27

$N(x)$ neighborhood of x, page 18

$o(H)$ size (or order) of schema H, page 41

P population, page 148

P_l link-specific bias, page 207

p_c crossover probability, page 147

p_i penalty parameter (GLS), page 137

p_m mutation probability, page 147

P_{acc} probability of accepting worse solutions in SA, page 95

P_{suc} success probability, page 189

R demand matrix, page 187

$r()$ random walk autocorrelation function, page 35

r_{ij} amount of traffic required between nodes i and j, page 187

s number of different decision alternatives, size of search space, page 19

s size of a tournament, page 149

S_f set of parameters that determine the objective value of a solution x (see $\mathscr{A}_X(x, S_X)$), page 25

S_X set of parameters that determine whether a solution x is feasible (see $\mathscr{A}_f(x, S_f)$), page 25

T tabu list, page 140

T terminal set (genetic programming), page 153

T tree, page 61

t_{conv} number of generations until a run terminates, page 189

tr_{ij} traffic flowing directly and indirectly over edge e_{ij}, page 187

$u_i()$ cost of solution feature i (GLS), page 137

V set of nodes or vertices, page 61

W distance matrix, page 187

w'_{ij} modified distance weights, page 207

$w(T)$ cost or weight of a tree T, page 187

w_j jth Walsh coefficient, page 40

w_{ij} distance weights in a graph, page 187

X set of solutions, search space, page 13

$x*$ optimal solution, page 32

x_g genotype of a solution, page 110

x_p phenotype of a solution, page 110

x_i decision variable, page 10

x vector of decision variables, page 13

Glossary

APX   constant-factor approximation, 89

BB   building block, 42
BNF   Backus-Naur form, 176

CV   characteristic vector, 188

EA   evolutionary algorithm, 147
EDA   estimation of distribution algorithm, 151
ES   evolution strategy, 142

FPTAS   fully polynomial-time approximation scheme, 89

GA   genetic algorithm, 147
GE   grammatical evolution, 175, 176
GLS   guided local search, 137
GP   genetic programming, 153
GRASP   greedy randomized adaptive search, 140
GS   greedy search, 64

HSS   heuristic space search, 166

ILS   iterated local search, 139

LB   link-biased, 207
LNB   link and node biased, 208
LP   linear programming, 50
LS   local search, 217

F. Rothlauf, *Design of Modern Heuristics*, Natural Computing Series,
DOI 10.1007/978-3-540-72962-4, © Springer-Verlag Berlin Heidelberg 2011

Index